THE FORTUNES OF THE WEST:

The Future of the Atlantic Nations

THEODORE GEIGER

The Fortunes of the West

The Future of the Atlantic Nations

γνῶθι σεαυτόν . . . καὶ ἡ ἀλήθεια ἐλευθερώσει ὑμᾶς.

INDIANA UNIVERSITY PRESS

Bloomington London

Published in Canada by Fitzhenry & Whiteside Limited,
Don Mills, Ontario

Library of Congress catalog card number: 72-76946
ISBN: 0-253-12710-6

MANUFACTURED IN THE UNITED STATES OF AMERICA

To

Nancy and Loren

Phina and Tom

who will help to shape the future

Contents

Preface

A PROFOUND TRANSFORMATION is now taking place within the nations of North America and Western Europe and in their relationships with one another. The purpose of this book is to present a way of thinking about this complex process that would be useful to policy makers and opinion leaders, as well as general readers, in interpreting the significance of the changes involved for their own interests and responsibilities.

Policy makers' tasks become more difficult from year to year. They must take into account an ever-diversifying and more complex range of interacting factors, whose probable effects they must also project over longer and longer periods of time. The new mathematical planning and decision-making techniques are certainly needed to cope with the steadily increasing volume and variety of relevant quantitative data. But, many of the most significant factors cannot be measured or even expressed in quantitative form. To be useful in policy making, information and ideas have to be analyzed in accordance with a comprehensive set of concepts that distinguishes the determinative variables, whether quantifiable or not, and interrelates them in ways which disclose both the limits of the possible and the possibilities for action. For this reason, the National Planning Association—which continuously provides governmental and private decision makers in the United States with a wide variety of quantitative projections and analyses—asked me to develop a multidisciplinary nonquantitative conceptual framework for public and private planning activities in the Atlantic region.

This book aims to carry out that assignment by identifying and interpreting the main trends shaping the future of the Western nations. Because many policy makers and opinion leaders are not experts in the social-science disciplines involved, it is written in nontechnical language. Where technical terms have been unavoidable, they are defined when first used. In the interest of readability, the footnotes are placed at the end of the book and the references are kept to a minimum. A thorough examination of the methodological and philosophical questions raised by the analysis is not attempted, although the most important of them are briefly discussed in Chapter I and in footnotes elsewhere to indicate the main outlines of the approach.

Thus, like my previous book, *The Conflicted Relationship: The West and the Transformation of Asia, Africa and Latin America* (New York: McGraw-Hill for the Council on Foreign Relations, 1967), the present work is not written with a scholarly readership specifically in mind. However, the unanticipated interest of many university scholars, and especially of their students, in *The Conflicted Relationship* encourages me to hope that they may find *The Fortunes of the West* similarly useful. As apparently the former book did with respect to the multidimensional complexities of the development process in Asia, Africa and Latin America, the latter may help to meet the need for integrating concepts and data by which scholars can relate their own specialized research to the other relevant aspects of the transformation of the nations of North America and Western Europe.

The interpretations and projections in this book reflect many years of study in the social sciences and related fields and the experience of service in and continuous subsequent contact with the United States government and international organizations. More particularly, they are based on over six hundred in-depth interviews with national-government and international-organization officials, businessmen, trade unionists, journalists, social scientists, students, and other opinion leaders during lengthy study visits in Western Europe in 1961, 1963, 1966, 1969 and 1971, in Japan in 1964 and 1967, in Australia in 1967, and in East Asia in 1971–72, as well as during frequent trips to Canada. I am most appreciative of the willingness of so many busy people to talk at length with me and, in many cases, to arrange interviews with others whose views they believed would be important for the study. I am especially indebted to Saburo Okita for the Japanese interviews and to John Thomas Smith and Sir John Crawford for those in Australia.

I am most grateful to the Ford Foundation, whose grant to the National Planning Association made possible the field studies from 1966 to 1970 and the other research and writing during those years. Naturally, I take full and sole responsibility for the interpretations presented here of the information and opinions given to me during the interviews and derived from other sources.

I am deeply indebted to friends and colleagues for their warm interest and extensive help: To Henri Aujac, *Directeur, Bureau d'Informations et de Prévisions Economiques,* who opened so many doors in France and whose insights into European developments and thoughtful comments on the draft of this book have been most valuable. To

Benjamin Nelson, Professor of History and Sociology, New School for
Social Research, whose contributions to the historical part of Chapter
II are so inadequately indicated by the references to his own pioneering
research. To my colleagues at the National Planning Association, John
Miller, Everard Munsey and Wilfred Lewis, Jr., and to my former col-
league Roger D. Hansen, for their continuous encouragement of the
work and helpful suggestions for revising the original draft.

I have also benefited greatly from the criticisms and comments of a
group of distinguished Americans, Canadians and Europeans: Max
Beloff, the late Waldemar Besson, Robert R. Bowie, Miriam Camps,
Fabio Luca Cavazza, Harold van B. Cleveland, Alphonse de Rosso,
Edward W. Doherty, François Duchêne, H. Edward English, Arthur A.
Hartman, Pierre Hassner, Klaus Knorr, Lawrence B. Krause, Mancur
Olson, John Pinder, Robert L. Rothstein, Andrew Shonfield, Altiero
Spinelli, Pierre Uri, George S. Vest and Richard D. Vine. I very much
appreciate the time and care they devoted to reading all or parts of the
first draft and to explaining their suggestions and disagreements.

Frances, my wife, has contributed as much to this book as I. She
participated in most of the interviews, undertook most of the library
research, edited the successive drafts, relieved me of as many of my
other responsibilities at the NPA as she could so that I would have time
to write, kept up my spirits when the ideas and the words were slow in
coming, and made all of the sacrifices that assured conditions harmonious with my peculiar working habits. As always, her rare capacity for
empathy with other viewpoints and other cultures and her good common sense have kept me from many misinterpretations and errors.

March 1972   Theodore Geiger
                              National Planning Association
                              Washington, D. C.

THE FORTUNES OF THE WEST:

The Future of the Atlantic Nations

What This Book Is About

FORETELLING THE FUTURE has always been a compelling human concern, and man's ingenuity has devised countless techniques for doing it, some cruel and many bizarre. Theologians and philosophers, as well as innumerable fables and stories, have warned in vain that people are usually better off not knowing what their fortunes will be. Nor has eagerness to foresee the future been diminished by the irony, portrayed by the great Greek dramatists, that, even when a person's fate can be prophesied, the hoped-for or feared developments usually occur in unrecognized ways, at unexpected times, and with unimagined consequences.

For all the paradoxical perplexities and practical difficulties involved, complex modern societies must try to forecast certain aspects of the future if they are to deal intelligently with their problems and meet the expectations of their people. In response to this need, methods more effective than divination have been devised for projecting the probable shapes and magnitudes of things to come. Yet the traditional limitations still apply. It is usually impossible to predict future events—that is, the specific forms in which expected developments will manifest themselves—for any considerable time ahead. But, like the Hebrew prophets who foresaw the impending consequences of existing injustices, modern forecasters can discern longer-term trends—that is, the ways in which, and in some respects the amounts by which, existing social characteristics and relationships are likely to change, granted certain assumptions about the determinative factors involved.

The urge to foresee future developments has become greater than ever in recent years. In part, this reflects the growing pressures on governmental and private institutions of all kinds to plan their activities over longer time periods and in terms of an increasing variety of interacting factors. In part, it expresses the rising anxieties of people generally, and of members of the elite groups in particular, about the survival of American society "as we know it" and the continued ability of the planet's biosphere to support human life. In response to both types of concerns, natural and social scientists, engineers and architects, politicians and publicists have been seeking to identify the factors shaping the future, explain how their influence can be projected, and prescribe means for controlling their effects. Indeed, so large and important has this forecasting and planning effort become that those engaged in it now have a name of their own: futurists.

In view of the growing volume of futurist writings, it may validly be asked: Why another book? How does this one differ from those already published?

Most of the futurist publications written specifically for the wider readership of policy makers and opinion leaders can be classified into two categories. Following Marx's famous distinction between the two kinds of socialism, they may be called "utopian and scientific futurism." The utopian futurists can in turn be divided into two types. Although they agree in their diagnosis of the forces shaping the future of American and other Western societies, their prescriptions for dealing with present and prospective problems are diametrically opposite. Both see science and technology as the main—indeed, to some the only—factors responsible for social and cultural development, good and bad. To the first type of utopian futurists, the rationalization of society through universal application of reason and science is the sovereign remedy for social and ecological problems, and they confidently predict that it will sooner or later inaugurate a new age of peace, plenty and progress for all mankind. To the second type, however, the causes of human and planetary ills are the monstrous institutions and ways of thinking resulting from runaway science and technology, which—if not totally rejected—are at least going to be reformed to serve humane values and human relationships through an imminent radical reconstruction of Western technological societies.

If the cause were as simple and the remedy as easy, quick and thorough as the rationalistic utopians claim, it is hard to understand why social conflicts and environmental deterioration have not already been overcome but, instead, seem so intractable. If the "Establishment" is as dehumanized by technological ways of thinking and acting and the people as corrupted by the pleasures of consumerism as the humanistic utopians insist, one wonders how their revolution could ever win enough support to succeed and, if it could, whether it would be better able than the French, Russian and Chinese Revolutions to preserve its purity and achieve its humane ideals. Both the causes of our difficulties and the possibilities for remedial action appear to me to be a great deal more complex, ambivalent and uncertain than they do to most utopian futurists. This is because my conceptions of how societies and cultures change and of how much and in what ways Western societies are likely to be transformed in the foreseeable future differ quite substantially from theirs.[1]

For example, technological innovations certainly will be major elements in social changes in the years to come. Yet, advances in science and technology do not happen of their own accord. Whether such developments occur, whether and how they are applied, indeed, their very shape and content, depend upon the character of the social systems in which they take place, the prevailing ideas about their meaning, value and social consequences, and the dominant concepts regarding the nature and purpose both of the physical universe and of man and society. Thus, the institutions, values and behavioral norms of a society determine its willingness and ability to undertake and apply scientific and technological developments, and importantly influence the kinds and directions of the innovations that it is capable of perceiving and conceiving.[2] And, as part of the ongoing social process, scientific and technological advances in turn affect both the capabilities of the society and its modes of seeing and thinking.

An effort, then, to understand the current transformation of American and other Western societies and to project its probable outcome must go deeper both in time and into the complexities of the sociocultural process than do either variety of utopian futurists. For example, with respect to the influence of science and technology today and in the future, the analysis must seek to explain why their extraordinary flourishing in the past four centuries has occurred only within Western civilization, what sociocultural forces have entered into the West's exceptional drive not only to understand nature and society but also to use its new knowledge to improve the conditions of life. For, what has distinguished Western civilization from the other great sociocultural traditions evolved on this planet are the institutions, values and modes of perception and conception that have molded and motivated its unparalleled plasticity and activism since the 11th century. The origins, development and present and prospective effects of these unique Western social and cultural characteristics—which in our own day have been exported to all regions of the world—are traced in Chapter II.

The second difference between the approach in this book and that of the utopian futurists relates to the nature and extent of the social changes likely to occur during the remainder of the century. It is my belief that the utopian futurists tend to exaggerate both. Even in societies as dynamic as those of the West, sociocultural continuity generally predominates over sociocultural change, else they would be too fluid and amorphous to perpetuate themselves. Chapter III will point out, for example, the persistence of traditional Russian social relationships, attitudes and behavioral norms amidst the many changes in political and economic institutions after the Bolshevik Revolution and the great increases in resource availabilities under the Soviet regime.

In the United States, the most dynamic of Western societies, an adult of the early 1900s would not feel that the conditions of the 1970s —of his third-generation descendants—were *fundamentally* alien to him. Virtually all of the differences that he would notice today were already underway or forecast in general terms (usually not, however, in their specific forms and actual timing) in his period. They would include both those that he might approve, such as technological progress and socioeconomic improvements in the conditions of life, and those he might deplore, such as the growing scale and pervasiveness of government activities, greater sexual freedom and equality, the decline of religious faith, nonrepresentational art. Hence, even if the pace of sociocultural development were to double over the next three decades, the world of the end of the century would probably not seem much more strange to the adult of today than contemporary America would to the adult of the beginning of the century. My own view of how much and in what ways American society might be changed in the 21st century is presented in Chapter VII.

If the utopian futurists exaggerate the extent of sociocultural change, the scientific futurists tend to underestimate it. Although this book is closer to the second category of futurist literature, it nevertheless differs from much of the latter in projecting a more fundamental transformation of American and other Western societies over the longer term. This difference arises in large part from the tendency of scientific futurists to concentrate on the development of methodology rather than on its application to forecasting sociocultural change and, when they do apply it, to deal only with those variables that can be expressed and projected quantitatively. These methodological and quantitative biases are understandable in view of their training either in the natural sciences, engineering, architecture and other disciplines concerned with physical phenomena, or in mathematics, operations research, linear programming, systems analysis, simulation and other methodologies involving abstract logical relationships and reasoning processes that are independent of particular subject matters.

In contrast, the analyses and projections in this book try to take into account not only the considerations of rational interest and the economic, demographic and other measurable trends shaping the future but also the wider social-institutional and the deeper psychocultural aspects of human decision making and action. When this is done, the determinative relationships are seen to be so diverse, complex and ambivalent, and to interact in so many crucial respects that are either unquantifiable or incommensurate that quantitative projections alone cannot reflect a large enough portion of the process as a whole to have much significance. Nor is it possible even to design a comprehensive mathematical model that could properly express the relative weights of the very numerous and different influences involved, adequately take into account their often contradictory directions and kinds of effects and their multiple interactions, and allow for the further complications resulting from their differential rates of change. For these reasons, the projections in this book are qualitative and generalized.

Moreover, this book endeavors to avoid in another way the tendency of scientific futurists to underestimate sociocultural change. It takes a much longer historical perspective than most of them customarily do. The quantitative methodologies commonly used by scientific futurists implicitly assume that the future is going to be like the present. Extrapolations of the probable size and rate of changes in the continuing trends are of necessity based on *current* data about the magnitudes and weights of the factors now operating that could produce modifications in the future. Thus, the stricter the adherence to econometric and other quantitative techniques, the more difficult it is to foresee really new developments resulting from genuine innovations in thought and behavior and novel recombinations of existing elements (as in dialectical processes and the transformation of quantitative accretions into qualitative changes). In highly dynamic societies like those of the West, such unforeseeable developments are highly probable.

It is possible in some cases to offset this inherent limitation of quantitative methodologies by studying the historical origins and past development of currently important trends. Information—both qualitative and quantitative—about the factors that in previous periods have not only sustained these trends but also introduced innovations into them can be helpful in suggesting analogous (not identical) or contrasting ways in which they might be transformed in the years to come. Admittedly, history can never be a clear or certain guide to the future. Nevertheless, utopian futurists, who deny that the past has any relevance to the future, fail to grasp that experience is the most reliable basis we have for modifying the behavior of societies, no less than of individuals, so that they can cope more effectively with present and prospective realities. For their part, scientific futurists often do not recognize that the modes and methodologies of rational analysis, no less than of unconscious motivation, are rooted in the developmental histories of societies, as well as of individuals. And, those who criticize all types of futurism, including mine, on the ground that they are capable of projecting only probabilities and not certainties forget that death alone is certain—not the most helpful assurance for dealing with the future problems of the living.

To perceive and grasp the significance of changes in the relevant persisting trends, therefore, this book traces their development over quite long periods of the past. As in a running broadjump, one has to start from a considerably greater distance behind in order to move very far forward. Hence, in some places, as in Chapter II, attention is devoted to the remote origins, as well as the recent history, of certain conceptions and expectations crucial to understanding the nature of the current transformation of Western societies and the roles played by leading nations in the international system. However, no attempt is made to present a comprehensive picture of the changing, economic, political and other sociocultural elements of past periods viewed in their forward developmental perspective, as in conventional historical studies. Rather, my analysis of past trends is retrospective and highly selective. Like Geoffrey Barraclough's method of "contemporary history," it seeks from the vantage point of the present to discern the evolution out of the past only of those comparatively few patterns of interrelated thoughts and actions that have become of major significance today and are likely to continue to be of determinative importance in the future.[3]

In addition to taking a longer historical perspective, this book has a wider geographical scope than many futurist writings.[4] Instead of concentrating exclusively or mainly on the United States, its focus is on the North Atlantic region, that is, North America and Western Europe.[5] The reason is that the nations of this region look to one another far more than to others elsewhere on the planet both for mutually valuable interrelationships of many kinds and for unfavorable effects against which they cannot insulate themselves. Their high degree of interdependence has three major aspects.

The first arises from the fact that the societies and cultures of North America and Western Europe have all developed and continue to exist within the same great historical tradition and contemporary milieu or framework known as Western civilization. True, each country differs from the others in significant respects. But, there are—and will for the foreseeable future be—a great many more similarities in their social structures and institutions, in their ways of thinking and of living and working, and in their material products and technologies than exist between them and the other varieties of Western civilization. And, the similarities are very much less between them and the non-Western societies and cultures in other parts of the world. In varying degree, all of the North American and West European countries are today experiencing the same kind of profound sociocultural transformation. Because of their basic affinities and the growing scale and rapidity of communication among them, each finds the greatest external stimulus to and constraint on its own transformation in its relationships with the others. Hence, in seeking to make the choices and meet the difficulties implicit in its own processes of change, each needs to understand how those likely to be experienced by the others would affect it for good and for ill in the years to come.

The second aspect relates to the economic interdependence among the nations of North America and Western Europe, as well as Japan, and to their collective importance in the world economic system as a whole. Today, they conduct three-quarters of their total external trade with and invest two-thirds of their private capital outflows in one another. These proportions reflect the needs and opportunities engendered by their own unparalleled economic development, as well as the fact that they have been deliberately reducing—and the European group has already abolished among its members—many restrictions on freedom of trade and capital movements. Not only do they all have a substantial interest in preserving and increasing their mutual commerce and investment but also the lowering of barriers has made each of their economies much more susceptible to the influence of conditions and trends in the others. Even the United States, although it is the least dependent on foreign trade owing to the continental size and diversity of its economy, has had to recognize that its own economic situation and policies are significantly responsive to changes in those of Western Europe, Japan and Canada. Moreover, because the Atlantic countries and Japan constitute by far the largest market for goods, capital and technology on the planet, they collectively exercise the preponderant influence on the world economic system. Hence, people in these nations concerned with the future economic structures, rates of growth, levels of employment, standards of living, stability of prices, technological advancement, and profitability of enterprise in their own countries need to take specifically into account the effects, favorable and unfavorable, of probable developments in the region as a whole.

The third aspect reflects the dangers perceived by the Atlantic countries in the world political situation. Although in process of change, world politics has hitherto been of such nature as to sustain the reciprocal convictions of West Europeans that their security depends fundamentally upon the United States, and of North Americans that, as twice before in this century, they will inevitably be involved in defending Western Europe against any attack on its freedom. Because modern military technology makes the issue of world peace and war more fateful in our time than ever before, the probable development of the international system and its effects on political and defense relationships among the Atlantic countries must loom large in the minds of policy makers and opinion leaders concerned about the future.

These intercultural and international relationships can also be expressed in another way, which helps to distinguish further the approach taken in this book. It is that each geographical level—namely, the Western nations individually, the Atlantic nations as a group, and the worldwide group of nations—can be conceived of as a system, that is, as an identifiable pattern of interrelationships whose development is in varying degrees both self-determined and influenced by its continuous interactions with the two other smaller or larger patterns.

In this conception, a national society consists of a geographically defined, self-governing collectivity of individuals and social groups more or less tightly patterned into multiple sets of continuing relationships (that is, organizations and institutions) which interact cooperatively and competitively in accordance with the cues and responses characteristic of the culture. Thus, neither institutions nor ways of thinking and behaving are alone determinative; one could not exist without the other, and they are continually reshaping each other at a greater or lesser rate of change. Nor, in this conception, can particular aspects of the society, such as the economy, or of the culture, such as science and technology, be regarded as determining the rest. They, too, are inextricably interdependent with the other aspects. However, at various periods in the development of a society, particular groups and institutions and particular modes of perceiving and conceiving may be more important determinative factors than others. Also, at various times, the national society may be more or less open to influences from outside, and may in turn have a greater or lesser impact on its regional system or on the international system as a whole. The larger-than-national systems are generally much simpler and more open to influences from their constituent nations than *vice versa.*[6]

This systems model of the social process and of international relationships may be described in a less formalistic and deterministic way. Just as individuals have certain physiological and emotional needs that are inherent in their biological and psychological constitutions so, too, do national societies and groups within them have particular interests that reflect their structures, the pressures and potentialities of their natural environments, and their situations relative to those of other nations and groups. Because needs and interests are capable of very wide diversity, they are always in some degree contradictory; and because the physical, emotional and social capacities to satisfy them are limited, they are always more or less in competition.

The resulting struggles within individuals, within societies, and among nations in greater or lesser degree involve the deliberate, conscious exploring of prerequisites and consequences and weighing of costs, benefits and risks. But, these rational operations are never the sole determinants of individual and social behavior. Other aspects of the process are also continually at work, largely at the unconscious level. The often conflicting physiological drives and emotional needs impel people to feel strongly enough about the opportunities they perceive to make the effort to protect and foster their personal, institutional and national interests. This psychic energy required to transform ideas into actions is responsible for the intensity with which people both cooperate and compete to satisfy their needs and interests. It is also expressed in the strength of their perceptions and conceptions of their interests and of their commitment to the kinds of behavior they regard as desirable and effective. Emotions thereby always in some degree reinforce, color, distort or displace the reality content of ideas and the rational motivation of actions.

The perceptions and conceptions of needs and interests and the forms in which emotions are felt and expressed reflect not only the nature of the realities involved but also two basic aspects of the learning process: socialization, in which people learn to relate in specific ways to the other persons in their society; and acculturation, in which they internalize the particular ideas, values and behavioral norms of their culture. Thus, in all societies, even the most primitive, children from the moment of birth begin to grasp consciously and unconsciously what and how they should see and not see, feel and not feel, believe and not believe, do and not do. These sets of socially approved cues and responses are rooted in the unique historical experiences, inherited traditions, and existing institutions and relationships of a society, and are adapted by each generation and transmitted by it to the next. This process by which societies and cultures are perpetuated is always influenced by changing realities, physical and social, and, in turn, it helps to change them.

Of special importance in shaping and orienting perceptions and conceptions are the deeply rooted and strongly felt rationales, values and aspirations, and the behavioral norms appropriate to them, by which a society justifies to its members—and to outsiders—its reason for being, its sense of purpose, its unique nature and destiny—the emotionally charged articulations that Benjamin Nelson has called its "dramatic design."[7] Until well into the 20th century, the dramatic designs of Western societies, as of all others, were central elements of the religious aspects of their cultures. Although the secularization process began during the Renaissance, explicitly agnostic and naturalistic ways of thinking about the nature and destiny of man and the universe became prominent among the intellectual elites only in the course of the 18th century, and they did not spread into mass education, popular politics, and the life-styles of the great majority of families at all social levels until after World War I. Even today in Atlantic countries, the originally religious character of their existing senses of national identity and purpose can still be discerned behind their now secularized terminology and their current validations by reason, history, racial superiority, or institutional or cultural preeminence, if no longer by the will of God.

The final characteristic of the approach needing a word of explanation is the fact that I do not offer specific suggestions for ways to avoid the harmful possibilities of the future and to enhance the probabilities of those believed to be desirable. In this book, I am not trying to influence policy directly but to satisfy one of the essential prerequisites for effective policy making.[8] Before solutions can be prescribed, the nature of the problem must be accurately diagnosed and the limits of the possible determined within which choices can be made. The effectiveness of policy decisions reflects not only the desirability of the goals sought and the appropriateness of the means selected for achieving them. It also depends on the acuteness of policy makers' understanding of the determinative elements in the ongoing life of the particular societies involved and the extent to which they can be influenced by deliberate actions. These limitations on policy makers vary in accordance with the characteristics of different periods. In the two decades after World War II, for example, the scope for U.S. policy making was broad, reflecting both the need to reconstitute a functioning international system and the disproportionate power of the United States relative to that of its allies and rivals. With the ending of the postwar period during the 1960s and the changes in the international system (analyzed in Chapter IV), however, the limits of the possible for American policy makers have substantially narrowed. Today and for the foreseeable future, they must have an even clearer perception of reality than in the postwar years and a much greater ability to distinguish the more from the less probable courses of development that would result from the interactions of conscious decision making with the other aspects of the sociocultural process.

This is the reason for the Greek quotation on the title page. The sentence "Know thyself and the truth will set you free" is a hybrid, joining the Hellenic injunction inscribed on the Temple of Apollo at Delphi with the Christian promise made in the *Gospel According to St. John* (VIII, 32). Apart from its religious meanings and its significance for analytical psychotherapies, the thought contained in the quotation has increasing relevance for contemporary policy makers, who have to be concerned about the realism of their conceptions not simply of the external social process but equally of their own inner motivations and expectations. As human capacity to understand and control nature and society has grown enormously in the course of the 20th century, those who exercise this power must be able and willing as never before to recognize the differences between interests and wish-fulfillments, between rational considerations and rationalizations, between socioeconomic projections and psychological projections. They need to be conscious of how the cultural characteristics and historical experiences of their society in greater or lesser degree distort and displace the reality content of their perceptions and interpretations of the situations with which they must deal. Such self-knowledge is not a panacea for assuring the practical effectiveness and moral validity of policy choices. It is, however, one of the indispensable requirements for enlarging the area of human freedom and improving the conditions of human life.

Because their influence on the Atlantic regional system is so great, the analysis in this book begins at the level of the national systems. The first part of Chapter II sketches in very broad outline the evolution of those ways of thinking and of those institutions that have been primarily responsible for the dynamism and activism of Western civilization, for its will and ability to master nature and improve society. The second part analyzes in more detail certain of the contemporary manifestations of these characteristics that are now motivating and shaping the national societies of the Atlantic region. Chapter III then applies this general analysis specifically to the Soviet Union and the United States—the two superpowers that have played the major roles in the international system since World War II—to assess how the relevant sociocultural factors within each of them are likely to affect their foreign policies and external actions in the decades ahead.

In Chapter IV, the analysis moves to the level of the international system as a whole, dealing with the only way by which developments outside the Atlantic region could decisively influence the future of the Western nations: a nuclear war between the superpowers. The chapter sketches the more and the less probable changes in the world political system that would have the greatest effect in determining the likelihood of this possibility.

The next two chapters focus on the Atlantic regional system *per se,* taking into account the influences surveyed in the preceding chapters that converge on it from the national systems of its leading members and from the worldwide system. Chapter V analyzes the problems and prospects of the movement toward European unification, the most important development experienced by these nations since World War II. Chapter VI explores the changing economic and political relationships and trends within the Atlantic region as a whole, and projects the more and the less probable ways in which they could develop over the coming decades.

Chapter VII returns to the national level of Chapter II, using the long-term developmental possibilities of the United States as the prototype for the other Atlantic nations. Resuming the analysis of changing institutions, values and behavioral norms, it presents two projections to define the extremes of the range of possible ways in which Western societies and cultures could evolve over the next generation or two. Then, it describes at greater length a median projection between the two extremes that has a higher probability and would represent as profound a change in Western civilization as did the Renaissance and Reformation and the subsequent scientific, industrial and democratic revolutions.

Trends in the Development of
the Emerging Technocratic Society

THIS CHAPTER traces in broad outline the origins, previous transformations and current manifestations of certain basic sociocultural trends in Western development. In their cultural aspect, they consist of the uniquely Western ideas and motivations I call "redemptive activism"—the moral imperative to work for social and individual improvement—and the related conviction that reason and science are the principal means for achieving these ends. In their social aspect, they are expressed in the gradual evolution of institutions organized and interrelated in increasingly rationalized ways in accordance with efficiency criteria. These trends are of central importance to our study not simply because they distinguish Western civilization from the other great sociocultural traditions evolved on this planet. More significant, in their contemporary forms, they are major factors shaping the changes now occurring within Western societies and provide clues to the possible new transformations in the decades to come.

Before sketching the historical development and current interactions of these trends, an important qualification must be made. Western civilization has been and continues to be the most dynamic of the great sociocultural traditions in large part because it is so complex, so rich in dramatic movements and countermovements, so full of variants, exceptions and contradictory trends. This fertile complexity and ambivalence in some degree elude even the most comprehensive and detailed efforts to describe Western development. How much more, then, must this qualification apply to the far more limited endeavor, attempted in this chapter, of abstracting from the immense historical perspective of Western civilization certain sociocultural strands and sketching their past evolution and current manifestations without excessive distortion or simplification. Nevertheless, the need to understand the depth and power of those elements in the Western heritage that so importantly help to mold the present and future of the Atlantic region outweighs both the difficulty and the risk.

The Origins of Western Redemptive Activism

Psychologically and historically, the concept of redemption is compounded of the rejection of death as final dissolution and extinction, and the yearning for a mode of being free of all the ills, sufferings and frustrations of human existence on earth. Virtually all societies have some notion of life after death, and many also expect that, for the deserving, it will be a blessed state of perfect harmony and complete fulfillment. But, whereas other societies have hoped for redemption *from* this world, Western society has also sought it *in* this world.

These two conflicting notions of salvation were first clearly formulated in Antiquity, reflecting differences in the Hebraic and Hellenistic components of Christianity. From the Hebrew prophets were derived the passion for social justice and the expectation that it would be realized in this life, in the Kingdom of God to be established on earth after the Messiah had defeated the evil oppressors. In contrast, the Hellenistic neoplatonic and gnostic strand in early Christianity insisted that redemption could only be *from* this world, which is essentially imperfect, transitory and, indeed, illusory in contrast to the changeless, perfect and eternal ideals—in Christian terms, the transcendental Kingdom of Heaven, beyond space and time, promised when Christ returns at the last days.

The sense of mission to spread the gospel—the "good news" of salvation—was also Hebraic in origin, derived from the Covenant on Mt. Sinai under which God chose Israel to be "a messenger unto the peoples, a light unto the nations." Universalized by the Hebrew prophets, this mission was by precept and example to show the way to that blessed age when "nation shall not lift up sword against nation, neither shall they learn war any more." Although for various reasons, Judaism was not a proselytizing religion in most periods, Christianity was from the beginning a missionary faith. The salvation granted by God's grace was open to everyone who "believed in the Lord Jesus Christ." Indeed, it was not expected to occur until Christianity had been preached to and accepted by all mankind.

With their hope fed by the apocalyptic prophecies of the *Book of Daniel* and the *Revelation of St. John the Divine,* many early Christians confidently anticipated the imminent return of Christ. He would convert the heathen, defeat the forces of evil under Antichrist, and then reign on earth for a thousand years of peace and plenty—the Millennium— after which the dead would be resurrected and the Last Judgment would separate those saved for the eternal bliss of Heaven from those damned to suffer forever in Hell. However, with the triumph of Christianity under the Roman Emperor Constantine, the transcendental neoplatonic and gnostic strand in Christianity became predominant. St. Augustine proclaimed as the orthodox doctrine that the reward for the suffering faithful comprising the "City of God" would be beyond space and time, and the millennial expectation of a perfected earthly kingdom was condemned as superstition. Nonetheless, preserved in the apocalyptic visions of both the Old and the New Testaments, expectations of the Millennium did not disappear. The approach of the year 1000 A.D.—the end of the first thousand years of Christianity—rekindled hopes of Christ's early return, and the crucial social changes that became manifest toward the end of the following century—the spread of commerce, the rise of towns, the revival of learning, the stimulus of the Crusades—reinvigorated the prophetic strand in Christianity.

The most influential expression of revived millennial ideas was in the apocalyptic writings of the 12th-century Cistercian abbot Joachim of Fiore and of his followers. Joachimite prophecies proclaimed three progressive ages of history culminating in the millennial kingdom, when both nature and mankind would achieve perfection. Denominated by the three persons of the Trinity, they were the past Age of the Father, of fear and servitude under the Mosaic Law of the Old Testament; the present Age of the Son, of faith and filial submission under the rule of the Church and the precepts of the New Testament; and the future Age of the Holy Spirit, of complete "love, joy and freedom; . . . when the knowledge of God would be revealed directly in the hearts of all men; . . . there would be no wealth or even property; . . . no [need to] work; . . . [and] no institutional authority of any kind."[1]

Derived from the Biblical description of the Garden of Eden and the tradition regarding the communistic organization of the early Christian Church, the Joachimite anarchistic conception of the millennial kingdom has exercised a profound, continuing influence, which— as we shall see—can still be felt today, albeit in secularized form. It has served as the model not only for redemptive prophecies and radical critiques but also for efforts to establish communities of Edenic innocence and plenty, especially during periods of rapid sociocultural change, as during the 17th and 19th centuries and again today.

Although the Augustinian view of redemption beyond space and time remained the orthodox doctrine of the Catholic Church, millennial expectations played a highly significant role in the motivations and activities of the missionary orders, especially the Franciscans and the Jesuits, in the settlement of the New World.[2] Messianic and millennial ideas have been even more widespread and important in Protestantism. Reinvigorated by Protestant reliance upon Biblical revelation, apocalyptic concepts of the Millennium provided imagery and rationales for attacking the Pope and the Catholic Church—identified with Antichrist and the forces of evil. Moreover, from the start of the Reformation to our own day, Joachimite ideas have provided the inspiration and the plans for the perfected religious communities actually established in the Old and New Worlds or prophetically forecast by many Protestant sects. In *The Tempest* (c. 1610), for example, Shakespeare has the "honest old counsellor" Gonzalo amuse the shipwrecked princes with an account of how he would govern that pictures the program goals of the radical Protestant sects in England and on the continent:

> *I' the commonwealth I would by contraries*
> *Execute all things; for no kind of traffic*
> *Would I admit; no name of magistrate;*
> *Letters should not be known; riches, poverty,*
> *The use of service, none; contract, succession,*
> *Bourn, bound of land, tilth, vineyard, none;*
> *No occupation; all men idle, all;*
> *And women too, but innocent and pure.*
> *All things in common nature should produce*
> *Without sweat or endeavour: treason, felony,*
> *Sword, pike, knife, gun, or need of any engine,*
> *Would I not have; but nature would bring forth,*
> *Of its own kind, all foison, all abundance,*
> *To feed my innocent people.*

Since the Renaissance, an immense variety of designs for reformed or perfected societies without apocalyptic and eschatological (i.e., end of the world) associations have been published by both lay and clerical social theorists. They range from purely hypothetical utopias (like that of Sir Thomas More, which gave this *genre* its name) to presumably more practicable blueprints for renovated commonwealths and small philadelphic communities. In these secularized conceptions, the anarchistic model of innocent spontaneity and miraculous abundance in a restored Garden of Eden is superseded by rationalized arrangements, usually patterned on Plato's *Republic,* for coping with the complexities and ambiguities of human nature and society.[3]

These changing conceptions were fostered by the increasing *this-worldly* orientation of the late medieval and early modern periods, which reflected the geographical expansion and growing complexity of Western society and its developing interest in understanding real-life problems and exploring possible remedies for them. And, this crucial transformation interacted more directly and powerfully with the second great sociocultural trend in Western civilization derived from Greek rationalism and individualism.

The Origins of Western Rationalism and Individualism

In our secularized contemporary society, the importance of reason and science is characteristically stressed. If the names of Joachim and other millennialists are now forgotten by all except scholars, we still remember and adequately appreciate today the monumental contributions of the Greeks to the development of the ability to understand and control the forces of nature and society. In consequence, there is no need to outline the origins and early evolution of rationalist and scientific ways of thinking. However, certain aspects of the Greek contribution important for our analysis are not as widely known and require more specific discussion.

A convenient starting point is the famous dictum of Protagoras, a philosopher of the mid-5th century B.C., that "[each] man is the measure of all things, of things that are that they are, and of things that are not that they are not." Here, in a single sentence, are the three unique characteristics of Greek rationalism and individualism. First, man himself is capable of grasping the nature of reality—of distinguishing between what can and cannot exist—without the need for divine revelation. Second, he can do so because of his reasoning capacity. Third, each individual—not man in the abstract—has this rational capability.[4]

Greek thought was predominantly rationalizing, that is, it strove to classify phenomena and ideas and to order them into comprehensive, internally consistent systems. This tendency to measure, compare and relate perceptions and conceptions in logical ways was manifested during the Graeco-Roman period in both intellectual and institutional forms. The major intellectual developments were the great naturalistic systems constructed by the leaders of the various philosophical schools

—Plato, Aristotle, Epicurus, Zeno and their successors. Implicit in these logically integrated world views was the idea of natural law, of invariant patterns of behavior inherent in the nature of things. The emergence of the concept of natural law was the intellectual precondition for the founding of such sciences as geometry, physics, astronomy, biology and medicine by the Hellenistic scientists.

With the spread of Christianity to the Greek world, the latter's rationalizing tendency was responsible for the elaboration and continuing refinement of a formal theological system for the new religion. This activity instituted a most important new characteristic in Western development. In all of the ancient religions, including Judaism, the existence of the world was explained simply by the inscrutable will or whim of a divine Creator. In contrast, the Graeco-Roman theologians wove a complex rational system into Christianity, which logically deduced the nature and destiny of the universe and man from the assumptions given by divine revelation regarding God's own nature and intentions. The persisting interest of Christianity in theological inquiry and disputation helped to keep the rationalizing tendency, and its logical methodology, alive even after philosophical and scientific activity ceased with the collapse of the Roman Empire and the onset of the "dark ages" of the early medieval period.

Greek rationalism was rooted in, and in turn helped to develop, the unique political and economic institutions of classical Greek society. The independent democratic city-state was an unprecedented form of *macro* social organization. The economic growth and expansion associated with it involved the development of larger-scale and more complex kinds of productive and commercial activities, including more rational types of business organization, management and record-keeping, than those previously carried on by the Phoenicians and other traders of the ancient Near East. Participatory democracy and intercity relationships fostered the skills of argumentation and persuasion. After the decline of the Greek city-states and the rise of Rome, the most important institutional manifestations of the rationalizing tendency were the efficient organization and procedures of the Roman imperial administration and the logical system of the Roman law.

Greek individualism reflected the emergence of the distinctive person from organic social groups, such as families, clans and tribes. In traditional societies, like those of preclassical Greece, a person's sense of identity is derived largely from the sociocultural characteristics of the organic social units to which he belongs rather than from his own personality traits or attainments. Moreover, traditional societies are ascriptive—the life careers and expectations of virtually all of their members are predominantly determined by the status into which they are born and not by their personal achievements. In the Greek city-states, the traditional social units were weakened and gradually dissolved while the new forms of political participation impelled people to think of themselves and to act as separate individuals. Released from the mutual responsibilities and loyalties binding together the older organic social groups, the individual perforce became self-making and self-responsible, able to experience new relationships and perspectives, to journey to new places, and to play new roles. Thus, the institutional changes in Greek society helped to create, and were in turn partly shaped by, the pressures impelling citizens to participate in the political affairs of their city-states and to act in their personal capacities as philosophers, scientists, explorers, merchants or soldiers without regard to kinship, class, status or the other restrictions of organic social units.[5]

Individualism was also fostered under the Roman Empire, despite its centralized autocracy. The Roman legal system defined in detail and universalized the rights and obligations of persons, as individuals, *vis-à-vis* one another and the emperor, as the embodiment of the state. And, in the doctrine of the Trinity, Christian theology explored at length the existential and moral meanings of the person. These developments were important not only in further articulating the idea of the separate and distinct person but also in preserving it during the early Middle Ages, when European society regressed into the decentralized, more organic and autarkic organization of manorialism and feudalism.

The later decades of the 11th century witnessed the inception of major social changes that, in the course of the 12th century, established the trends leading to the subsequent development of Western society and culture. Chief among them were the rise of towns—of the self-governing or partly autonomous urban communes—in Western Europe, and the economic expansion that both stimulated and was further accelerated by it. These changes created institutions and needs analogous to those in which Greek rationalism and individualism had originally emerged. At the same time, commercial relations with the Eastern Mediterranean and Moslem Spain, and the Crusades in those regions helped to revive knowledge of the achievements of Greek philosophy and science. The resulting renewal of the rationalizing tendency was especially marked in the independent Italian city-states under the stimulus of economic competition, the struggles among the different classes within the cities, and the wars among states. Social differentiation in the Italian city-states and the urban communes, and the weakening of feudal-manorial ties again encouraged the emergence of individuals. This process was greatly facilitated by the importance of the concept of person in Christian theology and the renewed interest in the universally applicable system of the Roman law. In consequence, individualization and social mobility in the late medieval period soon surpassed those in the Graeco-Roman world.

At the intellectual level, much of the rationalizing effort in the 12th and 13th centuries consisted of further enriching and refining the system of Christian theology with the help of revived Greek philosophical concepts and logical methods, as in the great systematic work of Thomas Aquinas. But, these activities also laid the foundations for the subsequent development of Western philosophy and science. Indeed, the evolution of modern scientific approaches to knowledge cannot be understood without the significant contributions of such medieval thinkers as Albertus Magnus, Robert Grosseteste, Roger Bacon and Nicholas of Cusa.

Until these late medieval centuries, the expectation of redemption in this world and the rationalizing tendency evolved mainly in parallel. However, beginning in the late Middle Ages, and increasingly from the 16th century on, they interacted with each other and with the related institutional changes. Three fusions resulting from these interactions are especially significant for our analysis.

The first, in brief, was the emergence within the millennialist tradition of the idea of progress. This is the belief initially in the possible and later in the inevitable improvement of nature and society, either concurrently or of one in consequence of the other. This optimistic, progressist conviction regarding man's future existence in this world was in marked contrast to the pessimistic and often retrogressist ideas that prevailed from Antiquity until early modern times. The Greeks held that "the golden age" was in the past, never to return. In the Roman Empire, Stoic and Epicurean ideas about cycles of improvement and decay—the turns of Fortune's wheel—predominated, sometimes including the expectation of eventual entropy, or running down, of nature and society. The dominant Augustinian Christianity, with its insistence on redemption *from* this inherently imperfect and transitory world, found these entropic ideas congenial, and the Renaissance humanists revived Stoic theories of historical cycles. However, from the 16th century on, the progressist aspect of Joachimite prophecies of the three ages of history, each better than its predecessor, was increasingly emphasized, and eventually became predominant in the 18th and 19th centuries.

The Rationalizing Effects of the Protestant Ethic and the Patrimonial State

More important than the expectation of progress in shaping the development of Western society were the other two fusions—discussed in this and the next section—of the interactions between redemptive activism and the rationalizing tendency. It was the genius of Max Weber to have discerned the essence of one of these fateful changes in ideas and institutions that emerged in the 16th century, gradually developed and spread during the ensuing three centuries, and became predominant in the Atlantic region in the past hundred years. Weber found it to be "the specific and peculiar rationalism of Western culture," and devoted much of his scholarly effort to studying this Western phenomenon and its differences of degree and of kind from the earlier rationalizing processes in Western civilization, as well as from those in the other sociocultural traditions.[6]

This "specific and peculiar rationalism" was the use of the criterion of functional relevance as the organizing principle of a life-style— that is, of the ways of living and working—of a new and increasingly influential social group. True, there was nothing new about the idea of functional relevance *per se*—that is, of insistence on a strict causal relationship between means and ends as a test of logical validity. Hitherto, however, it had been used as a tool of intellectual analysis, as in William of Occam's famous "razor," which enjoined elimination of all assumptions and hypotheses not absolutely necessary for the logical demonstration of a conclusion. Weber discerned the manifestation of this logical principle not only in the religious ideas of the Calvinist wing of the Protestant Reformation but more importantly in the mode of living and working of its most characteristic adherents.[7]

The theological aspect consisted of the merging of the doctrine of predestination with the distinctive logical elaboration by John Calvin and his followers of Martin Luther's concept of the "calling." Those predestined by God for redemption demonstrate their "election" and express their religious devotion by achievement in their vocations (that is, in the occupations to which God calls them), for by so doing they participate in the working out of the divine will for the governance and salvation of the world. Ministers in the Calvinist and derivative movements were no less hostile to the pursuit of economic gain as an egoistic goal and the exploitation of others to acquire wealth than were Catholic theologians and Lutheran and Anglican divines. But' they could see no better evidence by which a man would be certain that he was among "the elect of God" than the success of his voluntary effort to respond to God's call to action in this world. In turn, the likelihood of prospering in his vocation was directly proportional to the dedication and efficiency with which he worked. Hence, fulfillment of the Calvinist sense of mission to serve God required single-minded performance of the particular duties pertinent to one's occupation—in effect, the conscientious exclusion of other values, loyalties and satisfactions not directly related to the specific purposes and methods of a person's work.

The essence of the expression in the life-style of functional relevance is impersonal rational calculation in decision making: the dispassionate formulation of the practicable goals implicit in an activity, the deliberate development of increasingly efficient means for achieving them, and the diligent effort to foresee and, if possible, control enough of the internal and external variables to make the outcome reasonably certain. Such a rational approach necessitates eliminating all factors, sentiments and considerations that do not contribute significantly to the chosen objective. Weber called this Protestant austerity "in-the-world asceticism" and contrasted it with the "out-of-the world asceticism" of Catholicism and the other great religions.

> This in-the-world asceticism had a number of distinctive consequences not found in any other religion. [It] demanded of the believer, not celibacy, as in the case of the monk, but the avoidance of all erotic pleasure; not poverty, but the elimination of all idle and exploitative enjoyment of unearned wealth and income, and the avoidance of all feudalistic, sensuous ostentation of wealth; not the ascetic death-in-life of the cloister, but an alert, rationally controlled patterning of life, and the avoidance of all surrender to the beauty of the world, to art, or to one's own moods and emotions. The clear and uniform goal of this asceticism was the disciplining and methodical organization of conduct. Its typical representative was the "man of a vocation" or "professional," and its unique result was the rational organization of social relationships.[8]

In stressing the significance of the Protestant ethic of worldly asceticism, Weber had particularly in mind the importance of prevailing systems of moral valuation and regulation of conduct in fostering or discouraging various types of economic behavior. In medieval Catholicism, he observed:

> the wide chasm separating the inevitabilities of economic life from the Christian ideal . . . kept the most devout groups and all those with the most consistently developed ethics far from the life of trade. . . . The rise of a consistent, systematic, and ethically regulated mode of life in the economic domain was [incompatible] with the medieval institutional church's expedient of grading religious obligations according to religious charisma . . . and by [its] other expedient of granting dispensations. The fact that people with rigorous ethical standards simply could not take up a business career was not altered by the dispensation of indulgences, nor by the extremely lax principles of the Jesuit probabilistic ethics after the Counter-Reformation. A business career was only possible for those who were lax in their ethical thinking.
>
> The worldly asceticism of Protestantism first produced a capitalistic ethics, although unintentionally, for it opened the way to a career in business, especially for the most devout and ethically rigorous people.[9]

The inner certitude of righteousness and salvation and the self-confidence engendered by this religious sanction were especially important for the economic innovator of the 16th and 17th centuries, who was going against long-established customs in applying more productive technologies in mining and manufacturing, introducing more efficient methods of organizing the labor force, and initiating more aggressive marketing practices. As Weber pointed out, "nothing else could have given him the strength to overcome the innumerable obstacles, above all the infinitely more intensive work which is demanded of the modern entrepreneur. Furthermore, along with the clarity of vision [i.e., rational calculation] and ability to act, it is only by virtue of very definite and highly developed ethical qualities that it has been possible for him to command the absolutely indispensable confidence of his customers and workmen."[10]

Weber was careful to distinguish the new impersonal, rationalized form of capitalist organization and activity from the older types of capitalism. Antiquity and the later Middle Ages—and, to a lesser extent, China, India and the Moslem empires—had known daring merchant adventurers and hard-driving owners of large plantations and mines worked by slaves, great financiers and tax farmers, unscrupulous speculators and ruthless moneylenders, for all of whom control of substantial amounts of capital was essential to their activities. However, these older types of capitalist entrepreneurs were concerned not simply with the systematic increase of their wealth. They were also desirous of using it for personalistic goals unrelated to their business activities *per se*—the preservation and advancement of their families' social status, the achievement of noble rank and political power, conspicuous consumption and leisure, patronage of art and learning, support of religious institutions and charities, and civic contributions. There were many examples of such capitalists, typified by the Medici and the Fuggers, from the 12th century on, not only in Italy and Germany but also in France, England and the Low Countries. They continued to be the predominant form until the 18th century; and, indeed, certain characteristics of this older kind of capitalist persisted well into the 20th century, as explained in a later section.

In contrast, the new type of entrepreneur, intent upon conscientious performance of his vocation, was impelled by his ethical code to a life-style of work, sobriety, abstinence and probity, avoiding goals and activities unrelated to his business, and guiding his decisions, at home as well as in his office, by impersonal efficiency criteria. These attitudes and norms of behavior predisposed him to technological and managerial innovations. And, in fact, the period from the mid-16th to the mid-17th centuries, when the new type of entrepreneur emerged and began to spread, witnessed an accelerated rate of technological change in mining and manufacturing in Western Europe, and particularly in England, the stronghold of developing Puritanism. Contrary to the general impression, England experienced two industrial revolutions, not one—the much better known developments of 1750–1850 having been preceded by a smaller but nonetheless substantial transformation in the Elizabethan and early Stuart periods.[11] In both, entrepreneurs and workers imbued with the Protestant ethic—Puritan and other dissenters in the earlier revolution, Quakers, Methodists and other nonconformists in the later—played disproportionately large roles relative to their numbers in the society.

However, although the new, more dynamic form of rationalized economic organization and activity increased during the 17th and 18th centuries, it continued to be overshadowed by the older type of capitalist enterprise in England until the 19th century and in many parts of the continent until the 20th century. Its spread in the Catholic areas of continental Europe was so much slower owing in no small measure to the Church's continued uneasiness about the means by which profit is earned in commercial, industrial and financial undertakings and to the virtual exclusion of these activities from the traditional hierarchy of meritorious occupations.[12] Moreover, it was not only Catholicism that inhibited the expansion of the new, more rationalized form of enterprise. It was also discouraged by the established Protestant churches in England (Anglican) and on the continent (Lutheran) that were integrally associated with the growth of royal power in the consolidating dynastic states of the 16th through 18th centuries.

Weber regarded the dynastic state as the postmedieval European form of the general type of patrimonial state that exists in periods and civilizations in which the legitimate authority of a traditional social order extends effectively over a large number of local communities (e.g., villages, tribes, towns, estates).[13] Patrimonial states vary widely in complexity and sophistication, depending upon the nature of and relationships among the major social groups and institutions comprising the society and its technological level, economic productivity, and administrative efficiency. All have in common, however, rulership by a theoretically absolute central authority, who regards the resources— human and material—of the society as his by right of inheritance, and whose claim is validated by religious sanctions and immemorial customs. In practice, the autocratic ruler is generally restrained by the rights and privileges traditionally enjoyed by various groups and by the power of particular interests—feudal, urban and ecclesiastical— characteristically expressed in Western society through representative bodies (Estates, Parliaments, etc.).

From the late 15th to the late 18th centuries, the consolidating dynastic regimes in England, France, Spain, Austria, Prussia and the Scandinavian countries were among the most fully developed examples of the patrimonial state. Their rulers sought to abolish, replace with royal authority, or render purely ceremonial the hereditary offices and local powers of the great feudal magnates and lesser nobles, the rights won during medieval centuries by the urban communes and guilds, and the privileges and immunities of the Catholic hierarchy and religious orders (while in Protestant countries they made themselves heads of the new national churches). They endeavored also to regulate the operations of the other major institutional systems of the society, especially the economy and the schools and universities, so as to ensure that they served the internal and external aims of the ruling dynasty. Their control was reinforced by conspicuous construction and consumption, the inculcation of loyalty, the prestige derived from patronage of the arts and sciences, and the alleviation of acute public distress. The identity of national and dynastic interests in the patrimonial state was expressed in Louis XIV's famous declaration: *"L'Etat, c'est Moi!"* The ideal of patrimonial rule was the "benevolent despotism" of 18th-century political theory.

Actively or passively opposed by the feudal magnates, the patrimonial state came to depend upon new men, drawn principally from among the older type of mercantile and financial capitalists, who held office by virtue of royal favor and could be counted upon to be personally loyal to the king. And, just as little distinction was yet made between the king's private wealth and the public revenues, so the new officials of the patrimonial state regarded their private and public activities as coexistent and accumulated capital from both. Civil and military—usually also religious—offices of all kinds were sold, as were the contracts to collect designated taxes (tax farming); the grants of monopolies to manufacture, import or export particular commodities; and the rights to dispense licenses and other forms of official approvals of economic and legal transactions. These practices were designed to increase and make more calculable the revenues of the patrimonial state and to assure its control over the national economy, as well as to satisfy the interests of the rising class of bourgeois officials. In consequence, domestic politics consisted largely of the particularistic struggles for offices, monopolies, pensions and privileges among the contending families, factions and cliques into which both the feudal and the capitalistic elites were divided.

The economic policy of the patrimonial state is generally called "mercantilism."[14] Reflecting its pressing need for revenues to carry on its political, military, cultural and welfare activities, the patrimonial state sought to increase the taxable national wealth, which it identified with money (in that period primarily specie, i.e., gold and silver), by maintaining a favorable balance of trade and preventing the export of precious metals. To this end, it protected domestic agriculture and industry against import competition by tariffs and quotas, and fostered the export of manufactured goods through bilateral agreements with other nations. It restricted the use of foreign shipping and prohibited direct trade between its colonies and other countries. It provided subsidies and monopoly privileges to favored producers and distributors, and imposed various kinds of restrictions and regulations upon economic activity within its borders.

The growing diversity and scale of the patrimonial state's activities at home and abroad encouraged the systematization and routinization of its administrative institutions and procedures. Greatly expanding the practice begun in the medieval kingdoms, national policy making and administration became full-time occupations for many more officials both at the top executive level and in the lower ranks, which were increasingly differentiated into a hierarchy of grades with fixed salaries and purchase prices. Regular record-keeping, accounting and reporting methods were gradually introduced, and orders, instructions and interpretations were slowly standardized. On the continent, the rigorously logical system of the Roman law was revived to supersede the haphazard, incomplete and inconsistent body of feudal and customary law inherited from the medieval period. In England, the customary law was extensively revised and reinterpreted by the royal officials and courts to meet this need.[15]

Nonetheless, the rationalizing processes instituted by the patrimonial dynastic states were neither as rigorous nor as intense, neither as impersonal nor as austere, as the unique way of thinking and acting constituting the Protestant ethic. True, Machiavelli had instructed the prince in calculated techniques of policy making and implementation that would more effectively advance a ruler's interests; and the mercantilists had developed the "political arithmetic"—as Sir William Petty called it—of organizing and directing the national economy to serve the purposes of the patrimonial state. But these rationalizing processes embodied more heterogeneous and unrelated goals and methods, more inconsistent values and incongruent behavioral norms than those sanctioned by the worldly asceticism of the Protestant ethic.

In the latter, the relentless application of the behavioral equivalent of Occam's "razor" shaved away all ends and means not strictly relevant to the occupational activity *per se*. At the same time, the doctrine of predestination by a transcendental God infused self-interest with religious conviction in fostering the most conscientious performance of these functions. Puritan and other nonconformist entrepreneurs did not simply themselves abstain from all activities unrelated to the efficient pursuit of their vocations. They also vigorously condemned in others, and especially in patrimonial rulers and their supporters, the use of resources and time for advancing dynastic and territorial ambitions or family and individual status, and for luxurious living, ceremonial display, indiscriminate or compassionate charity, sports and amusements, even the support of art and science. Such pursuits of the kings and the courtiers, the officials and the older type of capitalist beneficiaries of the patrimonial state were deemed sinful because wasteful, and frivolous because functionally irrelevant.

The Calvinists' sense of righteous mission to labor in their occupations with rational self-control and single-minded efficiency, no less than the specific interests inherent in the economic activities to which they believed God called them, predisposed the members of these sects in England and on the continent to become the most stubborn enemies of the patrimonial state. Increasingly opposed to the doctrines and hierarchies of the established churches, whether Catholic, Lutheran or Anglican, the members of Calvinistic sects were equally against the political and economic policies of the patrimonial regimes in their countries. They condemned personal absolutism and royal centralization; official supervision and detailed regulations, monopolies and trade restrictions; the sale of offices; and other patrimonial practices. Their demands for religious independence, communal autonomy and freedom of enterprise led not only to the civil wars of the 16th and 17th centuries. Their religious convictions and the constraints of their situations also fostered the emergence of certain ideas and attitudes—especially the authority of the private conscience, the self-responsibility of the person, the sense of guilt over lack of vocational achievement, the right of revolution, the separation of church and state—that were major influences in the developing individualism and democracy of Western society.[16]

After the 17th century, when the harsh doctrine of predestination lost its hold over many of the original Calvinist sects, other theological concepts continued to sustain the Protestant ethic of worldly asceticism not only among them but also among the newer, increasingly important denominations—the Methodist, Baptist, Pietist and other evangelical movements—that rose to prominence in the 18th and 19th centuries in Europe and the United States. Moreover, validated by their results, the gospel of work, the morality of conscientious performance of occupational responsibilities, the logic of impersonal rational calculation, and decision making by strict efficiency criteria gradually spread far beyond the professing membership of the sects that had developed and nurtured them. Inculcated by innumerable practical moralists and popular rule books, these increasingly secularized norms of behavior came in time to express the "spirit of [modern] capitalism" and to contribute powerfully to the development of industrialism, a unique economic system destined to be decisively more productive than any previously evolved on the planet.

The Scientific Revolution and the Redemptive
Role of Reason

The third major fusion resulting from the interaction of redemptive activism and the rationalizing tendency was the emergence within the scientific revolution of the 16th and 17th centuries of confidence in the power of human reason not alone to understand nature and society, as among the Greeks, but also to reshape them for the improvement of life in the here and now. Like the origins of Western rationalism, the history of the scientific revolution is much better known today than that of the parallel sociocultural changes described in the preceding section. Nevertheless, the two aspects of the scientific revolution of greatest importance for the development of the redemptive role of reason tend often to be obscured or misinterpreted. The first was the substitution of empirical observation for divine revelation as the source of knowledge about the nature of the universe and of man. The second was the substitution of reason and science for divine grace as the agency by which the salvation of man and society would be achieved.

As Benjamin Nelson has explained,[17] the essence of the first change was the replacement of one kind of highly sophisticated rational system of ideas and behavioral norms by another. The struggle between religion and science that occurred during the 16th and 17th centuries essentially involved the confrontation of two *macro* rationales. Each used the methods of reason to integrate a consistent world view that purported to provide trustworthy knowledge of the nature and destiny of man and the universe. To the Church, reason served the essential function of elaborating the truths about the universe and man implicit in revelation, which was accepted on faith and could be validly interpreted only by religious authorities divinely ordained to do so. However, to scientists like Copernicus, Kepler and Galileo and philosophers like Francis Bacon, Descartes, Spinoza and Leibnitz, rational analysis could yield true knowledge only if it was derived from scientific observation and experimentation conducted by persons trained in the relevant disciplines and logical methods regardless of whether they were authorized interpreters of divine revelation.

Nelson points out that "the suave and flexible elite of the Ecclesiastical Establishment, who followed the works of scientists and philosophers with considerable interest, did not raise objections so long as the innovators made no inappropriate claims to truth or certitude which openly challenged received doctrine." But, he continues, the founders of modern science and rational philosophy were

> ... committed spokesmen of the new truths clearly proclaimed by the Book of Nature which, they supposed, revealed secrets to all who earnestly applied themselves in good faith and deciphered the signs so lavishly made available by the Author of Nature. Nature's Book, in their view, was written in numbers, never lied, whereas the Testaments were written in words, which were both easy and tempting to misconstrue. Men like Galileo and Descartes were vastly more certain about the truth *revealed* to them by number than they were about the interpretations placed upon Scriptures in the commentaries of theologians.

It was this conviction of the new scientists and rationalist philosophers that, in Nelson's words, *"objective certainty and inner certitude* were the indispensable signs of science, true philosophy, and just belief" which fostered the emergence of the redemptive role of reason —the second aspect of the scientific revolution important for current and future developments. In the worldly asceticism of the Protestant ethic, the elect of God were psychologically sustained by the inner certitude of their salvation demonstrated to them by the objective certainty of their vocational success. During the scientific revolution, the sense of rightness of their mission to discover the truth and the spectacular results of their work imbued the new scientists and philosophers with the self-assurance and courage required to persist in the face of the portentous opposition of the ecclesiastical authorities. This deep feeling of inner certitude and objective certainty was the psychological aspect of the intellectual conviction of the 17th- and 18th-century scientists and philosophers that empirical observation and rational analysis were the means for comprehending the intricate, self-regulating mechanisms of the divine plan in accordance with which God had created nature and man. And, in the course of the 19th century, when the Deist role of an Author of the Universe disappeared, human reason and empirical observation could finally be enthroned as the sole authoritative sources of knowledge.

Faith in the redemptive role of reason followed naturally from this conviction. To Catholic and Protestant theologians of the medieval and early modern periods, redemption either in or from this world results from God's grace, from the continuous workings of divine Providence in directing the great drama of human history on the cosmic stage that God created for it. However, following the revival of

Graeco-Roman concepts of natural law and their exciting demonstration first in the Copernican and later in the Newtonian systems, the day-to-day operations of divine Providence were seen to be natural phenomena, manifestations of the laws of nature fixed at the Creation for the governance of the universe. Hence, reason—which alone can comprehend them—must also be the instrument, in Descartes' famous words, for "making ourselves masters and possessors of nature."

With their faith in the intrinsic goodness of man, the 18th-century *philosophes* proclaimed that, by acting in accordance with the laws of "nature's God" and the dictates of "sovereign reason," man and society could be transformed both morally and institutionally. Finally, the 19th-century determinists—Hegelian, Marxian, Darwinian and Spencerian—insisted that such an outcome was not simply possible but inherent in the nature of the universe itself. Just as the will of God had formerly made the coming of the Millennium inevitable, so now the idealistic or materialistic dialectic of history, or the laws of natural selection or of social development guaranteed the eventual perfection of man and society either by abrupt revolutionary jumps or through gradual evolutionary progress.

The Emergence of Positivism and Marxism

Thus, by the opening decades of the 19th century, the key elements in the dramatic designs of contemporary Western societies had developed from the unique Western fusions of redemptive activism and the rationalizing tendency. They are: the certainty of salvation in this world and of the vocational mission to work for it, the commitment to functional relevance and efficiency, and the conviction that human reason and science are the infallible means for mastering nature and perfecting society. The amalgamation of these themes in various secularized forms was forged in the social pressures and psychological stimulations of the great institutional transformations of the period from 1750 to 1850, especially those in the economic system associated with the industrial revolution and in the political system with the French Revolution. In turn, these profound processes of social change were in part sustained and shaped by the ways of thinking and acting that resulted from these momentous fusions.

The term "technocratic positivism" is used in this book to designate the broad range of contemporary ideas and expectations stemming from these sociocultural developments. The word "positivism" was coined by Auguste Comte, and his systematic working out of the concepts involved was the most fully articulated and influential manifestation of them during the 19th century.

Comte saw history as determined by the progress of the human mind. He, too, in the tradition of Joachim of Fiore identified three progressive ages of history, each characterized by an increasingly realistic way of synthesizing a unified conception of nature and society. They were the theological, or fetishistic, age; the metaphysical age of abstract deductive thinking; and the modern scientific age of empirical, experimental and inductive ways of understanding the real world. In the scientific age, such *positive* knowledge will inevitably bring about great changes in society—for example, the industrial revolution—and Comte foresaw progressive improvements in the conditions of life.

But, as one of the founders of sociology—in fact, he named it—Comte was neither a utopian nor a simplistic rationalist. In contrast to many of his predecessors and followers, he did not believe that reason would soon create the perfect society. Instead, Comte stressed the importance of the nonrational determinants of human behavior—the psychological and social factors by which the successive syntheses of ideas about the world are formed and through which, in turn, these concepts produce their effects on the course of events. The key to hastening the predominance of scientific, or positive, ways of thinking and acting, Comte believed, is not the efficacy of rational techniques *per se* but the motivating power working through them of humanistic ethics founded on altruistic love. By altruistic love he meant not love of a particular person, group, society or country but love of the essential qualities that unite all humanity, past, present and future.

Thus, Comte held that an elite trained in the sciences and freed by its superior ethics from selfish personal, class and national interests would provide intellectual and moral leadership for reforming society and controlling the forces of nature for the good of all mankind. To help motivate and give moral guidance for the application of scientific knowledge to social reform, Comte formulated a *positive* religion, with its own doctrines and rituals, which had a considerable vogue in Latin America during the second half of the 19th century.

The earlier—if less systematic and profound—ideas of Claude Henri de Saint-Simon contributed significantly to the development of the technocratic aspect of positivistic ways of thinking. Saint-Simon recognized the immense potentialities of the industrial mode of production, and he was the first to grasp the importance and to project

the future development of the incipient professionalization of industrial management and technical personnel. These insights led Saint-
Simon to proclaim the industrial entrepreneurs and engineers to be
the new elite capable of transforming and perfecting society. For their
guidance, he formulated in his last and best known work, *Le Nouveau
Christianisme,* a renovated Christianity appropriate to the coming
industrial age of equality and plenty. In consequence of his emphasis
on the importance of industrialization and industrial experts—and
despite his socialism—Saint-Simon attracted a considerable following
among businessmen, engineers, economists and technicians, who
further developed and widely publicized his ideas after his death.

Although Comte and Saint-Simon had stressed the central role of
moral values in motivating and guiding the new elite, their followers
tended increasingly to favor rationalist rather than ethical prescriptions for ensuring social perfection. As a result, the subsequent development of the wide range of positivistic attitudes and ideas and
the various forms in which they are manifested today all have in
common the implicit or explicit assumption that the problems of man
and society will sooner or later be solved by planned human action
made effective by the power of reason and the efficiency of functional
relevance. A self-chosen and self-perpetuating elite, qualified by its
rationalism and expert knowledge and legitimized by its increasingly
successful mastery over nature and society, has the mission of bringing
into being under its naturally beneficent rule a scientific age of peace,
plenty and limitless intellectual and material progress.[18]

In this most general sense, contemporary positivism embraces a
broad continuum of activistic, optimistic, progressist ways of thinking
and behaving. They range from realistic efforts at social reform and
individual improvement, at one end, to enthusiastic attempts to carry
out the latest panaceas for resolving the perplexities of human nature
and eliminating the deficiencies of society, at the other. Positivists
characteristically advocate a greater or lesser degree of rationally
planned action by centralized state authorities. Government planning
is needed, they believe, to supersede or to complement the effects of
the competitive pursuit by individuals and groups of their rational
interests, and to guide and accelerate toward chosen goals the progress
naturally resulting from evolutionary social processes. There is even
an exceptional democratic form of positivism—the unique American
common-sense popular positivism described in the next chapter.
Nevertheless, by virtue of the crucial role assigned to the scientifically

trained, self-selecting elite, positivists tend toward a presumed benevolent authoritarianism which, in the more democratic societies of
North America and northern Europe, is usually implicit rather than
overt.

Among the many varieties of positivistic ways of thinking that
emerged in the 19th century, one of the most distinctive, important
and long-lived has been the set of ideas and expectations designated
as Marxism. Although Marxism soon regarded itself as in opposition
to the mainstream of positivism, both sprang from the same cultural
tradition of redemptive activism and rationalism. However, each gave
greater emphasis to one or the other of the two guarantees of the
inevitability of progress and eventual perfection—the efficacy of
reason and science in the case of positivism, the laws of historical
materialism in that of Marxism.

Marxism has always been constrained officially to stress historical
determinism in consequence of Marx's reduction of social causation
to changes in the relationships, or mode, of production. Nonetheless,
in practice, Marxists have had to concede more and more efficacy to
the "ideological superstructure" in order to account for the leading intellectual and organizational role of the Communist Party, the insistence upon the importance of doctrinal orthodoxy, and the efforts
devoted to agitation and propaganda. Through the class struggle, the
dialectic of history has operated to bring about the three progressive
stages of ancient slavery, medieval feudalism, and modern capitalism,
and will terminate in the next and final historical period of the perfected classless society. The industrial proletariat created in the stage
of capitalism and organized and directed by a dedicated, disciplined
elite—the Communist Party—is chosen by history for the world-
redemptive mission of overthrowing in the "final conflict" the capitalist oppressors and their state—"the executive committee of the
exploiting class." After a transitional period during which property
and exploitation would be abolished, the ensuing personal freedom
and limitless scope for individual development would end the age-old
alienation of man from his own true nature and the products of his
labor, universal peace and material plenty would prevail, and the
state and other forms of institutionalized authority would "wither
away." Although the totalitarian regimes of existing communist nations are the antithesis of such expectations, this modern descendant
of the Joachimite vision continues in a variety of competing Marxist
versions to give meaning and direction to large numbers of people in
all parts of the world.

Marxism reveals its original roots in religious millennialism much more clearly and directly than does positivism. The Marxist version of human history, past and future, is as explicitly eschatological and messianic as that of the *Book of Revelation,* although the catastrophes are now social rather than physical, the forces of evil are capitalistic rather than satanic, the final battle is worldwide and not at Armageddon, and the Messiah is a German philosopher or a Russian dictator and not the Son of God. Indeed, particularly in its contemporary Russian and Chinese forms, Marxism can be considered a semi-religion, and this characteristic contributes to its open and unremitting hostility to the full religions, which it regards as rivals. (In contrast, secularized positivism tends to ignore religion.) Marxism is a semi-religion not simply because of its apocalyptic imagery but more importantly because of its faith in the dialectic of history as the guarantor of final victory, its fanaticism in justifying its means solely by its ends, its enforcement of doctrinal orthodoxy, its suppression of heresy and dissent, and its missionary zeal. The Marxist conception of the classless society of anarchic innocence and personal fulfillment is closer to the original millennial model than are the leading positivistic plans for the perfectly rationalized society. Marx and Engels scornfully denounced Charles Fourier and Saint-Simon as "utopian socialists" but the fact is that the designs of the latter two were far more sophisticated and their efforts to grapple with the perplexities of social organization and human psychology much more systematic than the fragmentary pronouncements about the coming communist society of the self-proclaimed fathers of "scientific socialism."[19]

Institutional Aspects of the Rise of Technocratic Positivism

Although the distinctive ideas and expectations subsumed in the positivistic strand of Western culture were fully articulated by the mid-19th century, they did not become the predominant way of thinking and acting of the elite groups in Western Europe and North America until the 1960s. Attainment of their present preeminence was dependent upon certain social changes of the past hundred years. Of special importance was the acceleration of these developments during the 20th century by the cumulative pressures and expectations of the great depression of the 1930s, World War II and its aftermath in the 1940s, and the cold war of the 1950s. These social changes occurred both in particular economic and political institutions and in the more general system of order and meaning of Western societies as a whole.

The beginnings of the rationalization of economic organizations can be discerned as early as the 12th century but this process reached full development and preponderant importance only in the mid-20th century with the emergence of the large modern corporation. It is distinguished from earlier forms by the fact that managerial control is divorced from ownership and is exercised by officials who qualify for their positions by specialized training and objectively measured career performance. In contrast, the management of the owner-operated family firm—which had for many centuries been the predominant form of organizing economic activities among the older type of capitalists and the Calvinist entrepreneurs alike—was determined by the inheritance of property rights by successive generations, their capacity for "on-the-job" learning, and their entrepreneurial vigor.

The difference between the modern and the older organizational forms can be viewed in several significant perspectives. One relates to the conception of the nature and purpose of the organization in the minds of those controlling it. With the gradual secularization of the Protestant ethic in the course of the 19th century, there was an accompanying relaxation of worldly asceticism in the life-styles of many —probably most—of the heirs of the founders of family firms. Attracted by the social prestige and hedonistic gratifications of the traditional elite groups, succeeding generations of owner-managers tended more and more to regard the family firm as a means for assuring the income needed to sustain aristocratic or *nouveau-riche* life-styles. In contrast, the large modern corporation has increasingly been conceived by its managers as an end in itself, an entity with an existence of its own, impelled to maintain its profitability, competitive position and business prestige.

Another perspective relates to the much more rationalized and impersonal character of the modern management-run corporation compared with the older family-dominated firm. This difference results from the corporation's more pervasive application of the principle of functional relevance and its much greater professionalization.

The worldly asceticism of Calvinist entrepreneurs was expressed in the rational calculation by which they conducted their economic activities, and it helped to produce the notable advances in productivity associated with their emergence. Insistence upon the strict relevance of ends and means continued to be a determinative characteristic of family firms in succeeding centuries. Yet, while careful of expenditures, obsessed with avoiding waste, and determined to extract the last possible hour of labor from the work force, 19th-century industrialists were only dimly aware of the gains in productivity that could be achieved by systematically applying efficiency criteria to every aspect of their businesses. It was not until the rise of the modern corporation in the 20th century that more and more of the functions and divisions of the enterprise began to be studied scientifically: first the work process, its steps, motions and physical layout and environment, and its duration and psychological preconditions and effects; then the employment, training and supervision of labor, and the recruitment, qualification and promotion of executive and technical personnel; next the methods of determining and paying wages, salaries and employee benefits; then the raising of capital and the management of cash flows; next the planning of long-range marketing, investment and inventory strategies; and finally the external relations of the corporation not only with suppliers, customers and competitors but also with the government, the communities in which it operates, and the public generally. This process began in the United States and has been much more widely and rigorously applied by American corporations than by their European counterparts for reasons discussed in the next section.

The practice of subjecting every aspect of corporate activities to the test of efficiency criteria was dependent upon, and hence helped to stimulate, the professionalization of managerial and technical personnel. Writing in the early years of the present century, Max Weber stressed the crucial significance

> . . . of the trained official, the pillar of both the modern State and of the economic life of the West. He forms a type of which there have heretofore only been suggestions, which have never remotely approached its present importance for the social order. Of course the official, even the specialized official, is a very old constituent of the most various societies. But no country and no age has ever experienced, in the same sense as the modern Occident, the absolute and complete dependence of its whole existence, of the political, technical, and economic conditions of its life, on a specially trained *organization* of officials.[20]

Weber had mainly in mind the legally trained official, who then predominated in governmental institutions, especially in Europe, and the technically trained engineer, whose future importance had long before been discerned by Saint-Simon. Beginning in the 1930s, however, two other kinds of trained officials have become even more important—the first in power and the second in influence—not only in the economic system but in the political and other major institutional systems as well. The former consists of the executives of business firms and government agencies—and increasingly today of other types of organizations—specifically trained in the new administrative principles and methods. The latter is comprised of the economists, sociologists, psychologists, mathematicians, physicists and other scientists who are applying their specialized knowledge to policy formation, program planning and the development of more sophisticated decision-making techniques in the main institutional systems of Western societies. Along with the engineers, the two newer groups constitute the modern technocrats *par excellence*.

The emergence of these practitioners of technocratic positivism has naturally gone hand-in-hand with the development of the relevant scientific and technological disciplines. And, this relationship, too, has largely been a phenomenon of the middle decades of the 20th century.

In previous centuries, the institutional and intellectual interactions between the physical and social sciences, on the one hand, and their applied technologies, on the other, were on the whole unplanned, indirect and slow. True, significant governmental support for technological development began in the 15th century, but it was mainly for military and naval purposes until the 19th century. Conscious, extensive, close and productive interdependencies between science and technology, and between scientific and technical research personnel, on the one hand, and government and business organizations, on the other, have gradually developed only over the past hundred years— the deliberate stimulation of technological advances by business firms only since the mid-19th century and the deliberate effort to interrelate pure science and technological innovation briefly during World War I and continuously only since the eve of World War II.

The new administrative and decision-making techniques began to be fostered by and applied in large corporations only after the turn of the century, when they also started to use some of the results of research by economists, industrial sociologists and psychologists. Governments commenced to adopt modern management methods and to utilize trained economists and other social scientists only under the spur of World War I and, more particularly, of the great depression of the 1930s. Nor was economics able to provide concepts and prescriptions for managing economic systems as a whole until the post-World War II acceptance and further development of the ideas stemming from the work of John Maynard Keynes and his Swedish precursors during the interwar years. All of these trends reached decisive proportions only in the 1950s under the stimulus of the cold war and the acceleration of the other pervasive changes in Western societies sketched in the next section.

The deliberate application of science and technology in the activities of economic and governmental organizations has also evolved in continuous interaction with the development of institutions designed to provide the requisite professional training and to carry on the necessary technical research. They were of two types. The first included schools of technology and engineering, of business and public administration, and of other specialized disciplines established within universities. The second consisted of independent research institutes and laboratories working in the physical and social sciences. Until the 1950s, the former type was almost exclusively a North American development, while Western Europe relied mainly upon the latter. Much of the intellectual inspiration and systematic methodology of both types were derived from German advances in the applied physical sciences and in the social sciences generally during the second half of the 19th century.

With respect to economic, governmental and other organizations in the Atlantic countries, the process of technocratization has not yet reached its fullest development. Even in the United States, where the transformation has been more rapid and pervasive than elsewhere in the Atlantic region, most of the officials in the upper levels of government agencies, large business corporations and other institutions received their education prior to World War II. Hence, most of them lack advanced or specialized training, which was not generally required during the interwar and immediate postwar years for initial employment or subsequent career promotion. Indeed, it was only in the 1950s that American corporations and other organizations began to recruit a majority of their new managerial and technical personnel from among the graduates of the schools of business administration, technology and engineering, and physical and social sciences. Still at middle-management levels, the younger executives and technicians most thoroughly imbued with positivistic expectations and norms of behavior, and highly trained in technocratic skills, will not become the dominant decision makers until the end of the 1970s in the United States and even later in Canada and Western Europe.

From the Patrimonial to the Technocratic Order

These changes in the organizations comprising the main institutional systems of Atlantic nations were a part of the general transformation of Western societies as a whole that has been underway in the course of the 20th century and reached decisive proportions in the 1960s. In its all-embracing, or *macro* manifestation, this transformation involved changes in the nature of the social order and of its dramatic design, its sense of identity and purpose. Both the institutional and the cultural aspects are subsumed in the contemporary concept of the nation-state, a distinctive form of *macro* social organization that has been gradually evolving in Western civilization since the 11th century.

Unfortunately, the term "nation-state" is not susceptible to precise and universally accepted definition. Its ambiguity reflects the different aspects of the complex phenomenon it designates. The nation-state is a set of institutional systems—that is, a determinate social order composed of continuing patterns of interactions, of cues and responses, among the individuals, groups, organizations and larger institutions existing in a geographically delimited and usually politically independent collectivity with some kind of sovereign authority. However, although it partly manifests itself in the political and administrative organizations of the national society, especially in those of the central authority, it is not the same thing as these institutional systems. The nation-state is also a cultural system—that is, a persisting set of feelings, perceptions and conceptions relating to the sense of identity and purpose of the national society as a whole. The self-conceptions, values and loyalties subsumed in the way of thinking called "nationalism" are generally considered to be its most fully developed and conspicuous manifestation. However, there were earlier forms which also played vital roles in the development of the universalistic values and norms of behavior that help to offset the divisiveness of particularistic interests and maintain the cohesiveness, orderliness and minimum consensus needed for an effectively functioning society.

This intrinsic vagueness helps to account for the different Western conceptions of the nature and significance of the nation-state. On the one hand, it provides room for Hegelian and other mystical reifications of the state as the manifestation of a divine intelligence immanent in history and of the nation as the embodiment of the biological or spiritual genius of a people. On the other hand, it is susceptible to reductionist definitions, such as the class-conspiratorial notion of Marxism (Engels' "an organization of the possessing class for its protection against the nonpossessing class"), and the more sophisticated Anglo-American pluralistic conceptions that equate the state with neutral agencies of government and regard it as deriving its form and functions at any particular time from competition and bargaining among the various interest groups comprising the nation. That the nation-state is something less than the presence of a transcendental spirit and something more than a colorless official bureaucracy is implicit in its dual institutional-cultural character. But, what it may be has so far defied more precise definition.

Many of the distinctive features of the contemporary nation-state were already manifested in the evolving patrimonial dynastic states of the 15th through 18th centuries. However, for the analysis here, the differences are more important in defining the long-term sociocultural trends that have been among the major determinants of the existing characteristics of Atlantic nations and will continue to help shape their future development. The major significant changes can be traced from the *patrimonial order* of the early modern period, through the intervening *liberal order* of the 19th and early 20th centuries, to the *technocratic order* now rapidly emerging in all of the Atlantic countries. Each is described in its pure form—as one of Max Weber's "ideal types"—to bring out its distinctive elements, but some of its actual variations will be briefly noted as well.

The social order and dramatic design of the 19th-century liberal state contrasted markedly with those of the patrimonial form, described earlier in this chapter, largely because it emerged in opposition to its predecessor. Although reaching in certain respects its fullest expression in the United States, the liberal order evolved in Great Britain in consequence of distinctively British political experiences, economic pressures, and religious and philosophical conceptions. Its main characteristics were shaped by the 17th-century revolutions against the patrimonial regime of the Stuart dynasty, by the struggles of dissenting and nonconformist entrepreneurs against mercantilist restrictions and political discrimination, and by the rationalism of John Locke and the utilitarianism of Jeremy Bentham and their followers. The United States, Canada, Australia and New Zealand were predisposed toward developing the institutions and values of the liberal order by their background of British settlement, their continuing ties with the mother country, and the absence of prior-existing patrimonial regimes on their territories. In contrast, a liberal social order and dramatic design only partially emerged on the European continent, even less in the predominantly Catholic countries than in the Protestant ones.

Nor was it an accident that the liberal state reached its fullest development in those countries most heavily influenced by Calvinist and derivative sects. For, in essence, it represented a secularized manifestation of the Protestant ethic of worldly asceticism, no longer expressed only in the life-style and business activity of the newer type of entrepreneurs but now also in the purposes and system of order of the society as a whole. In effect, Occam's "razor" was applied to the political regime: both the goals and the functions of government in liberal states were drastically pruned. Ideally, it was believed, the purposes of administration should be limited to preserving national security and domestic order, dispensing justice, and collecting the minimum revenues necessary for these functions.

In accordance with this conception, the liberal states abolished virtually all domestic restrictions on private economic activity, and—although Great Britain alone adopted free trade—many nontariff controls on foreign commerce and on capital and specie movements were also eased or eliminated. For the deliberate regulation of the economic system by a central authority, the liberal states substituted the rule of the natural laws of society—the "invisible hand" of market forces—whose unrestricted operation would automatically result in "the greatest good for the greatest number."[21] In this secularized version of the divine call to action in this world, the pursuit of their self-interests by individuals and organizations in a rational, impersonal manner was believed to be the best—indeed, the sole—way of advancing the interests of the society as a whole.

Thus, the liberal state did not depend for its effective functioning and for the steady progress of society upon the initiative of a benevolent, all-powerful central authority. Instead, it expected improvement to result from the rigorous and voluntary application of the principle of functional relevance by self-instigating and self-responsible individuals dispersed throughout the society. This decentralized individualistic social order was validated not by the authority of the past but—as in the Calvinist use of vocational success as the evidence of salvation—by the pragmatic test of present utility, that is, of how much pleasure it produced and how much pain it eliminated or avoided (Bentham's "calculus of pleasures and pains" or "system of moral arithmetic"). In this way, a measurable scale of benefits served as an objective and universal standard for decision making.

In the continental European states, the French Revolution and the Napoleonic conquest inaugurated rapid changes in the patrimonial dynastic order that, in the course of the 19th century, substantially reduced, if they did not everywhere abolish, royal absolutism and mercantilist restrictions on private activity. Nevertheless, a much greater scale and diversity of governmental functions persisted on the continent, and the state was still expected to provide the main initiative, ideas and resources for social progress. Except for Great Britain, therefore, the European countries continued to regulate and stimulate their economic systems and to restrict their foreign trade to serve transcendent national goals, as well as private interests.

Thus, in both their institutional and their cultural aspects, the continental European states preserved into the 20th century much more of the patrimonial social order and dramatic design than did the English-speaking countries. On the continent, this continuity facilitated the adoption of the new governmental functions and techniques required to meet the increasing pressures and rising expectations of the 20th century. But, until the 1960s, the persistence of major elements of the patrimonial order and dramatic design also inhibited the rationalizing and professionalizing of organizations and the deliberate development and application of new technologies, administrative methods and decision-making techniques in the private sectors. The opposite effects were experienced in greater or lesser degree by the English-speaking nations, in which the liberal order was most fully developed. In the United States, for example, the much smaller scope and diversity of government functions and the rationalist individualism—as well as other factors discussed in the next chapter—powerfully fostered the technocratization of private economic organizations and strengthened both elitist and popular faith in the redemptive efficacy of reason and science. Yet, the *laissez-faire* conception of the liberal state delayed for decades the assumption by the government of new responsibilities and functions for improving society.

The onset of changes leading toward the contemporary technocratic order is discernible in the second half of the 19th century. On the institutional side, the rising productivity of the industrial system began to provide resources both for unprecedented increases in consumption by a growing population and for the realization of a broadening range of interests and aspirations—individual, group and national—previously beyond human competence.[22] On the cultural side, the spreading positivistic faith in the power of reason and science, the sense of redemptive mission, and the expectation of continued social progress and eventual perfection were stimulated by and in turn helped to foster the new economic and technological capabilities. Together, these interactions generated the slowly growing conviction that the new resources and skills must be used to realize the values of justice, equality and welfare, which had hitherto been regarded as ideals to be achieved, if ever, in some distant future.

The gradual recognition in practice that industrialization *per se* would neither inevitably fulfill these expectations nor even automatically mitigate the sufferings of workers and others adversely affected by it led to rising pressures for remedial actions. Two types of governmental responses in the 19th century began the movement away from the functional austerity and *laissez-faire* policies of the liberal state. Predisposed to a minimal passive role for the government, the British initially adopted the regulatory approach, exemplified first by the Factory Acts and later by additional rules for relieving the hardships of industrial and agricultural workers and protecting them against dangerous and unhealthy working conditions. On the continent, the persistence of patrimonial institutions and paternalistic attitudes was conducive to the more active and directive social-welfare approach, most fully expressed in Bismarck's pioneering system of accident, sickness and retirement benefits. By the turn of the century, however, the leading European nations were employing both types of measures. For reasons already indicated, the United States lagged behind Europe in undertaking central government actions to ameliorate the adverse effects of industrialization, resorting to federal regulatory measures only toward the end of the 19th century and delaying the inception of national social-welfare programs until the depression of the 1930s.

These trends were accelerated and their scope broadened in each succeeding decade of the 20th century. With respect to the realization of social values, the prosperity of the first decade of the century and of the 1920s no less than the sufferings of the great depression and the promises of postwar benefits during the two world wars stimulated successively greater expectations. By the mid-20th century, it was generally accepted on both sides of the Atlantic that the rate of economic growth, the level of employment, the standard of living, and the distribution of income could no longer be left to determination by market forces alone, which now had to be supplemented and guided by deliberate policy choices and effective government actions. Also, in the course of the present century, this change was further accelerated by the measures that governments were compelled to undertake to meet the necessities of the two world wars and their recovery periods, and of the cold war, which involved both the costly armaments race with the Soviet Union and the political and economic efforts to prevent the spread of communism in various parts of the world. All of these developments perforce fostered technological innovation and improved the ability of governments to manage their economic systems so as to achieve defined goals.

Finally, in the course of the 1960s, the scope and diversity of the social values and national goals to be realized as rapidly as possible expanded still further, along with the prospect of ever-continuing increases in the already immense productivity of highly industrialized economic systems. Today, in addition to their many previously acquired functions, governments seek to provide minimum satisfactory incomes and equal opportunities to all, assure rising standards of education and health, protect and improve the physical environment, rebuild the cities, foster and finance the advancement of knowledge, support the arts, expand recreational facilities to meet greater leisure and earlier retirement, and in a growing variety of other ways better the quality of life for an increasing population. These new goals and expectations are being met not only by enlarging the range and diversity of the public sector but also by enlisting, pressuring and regulating the private sector. In varying degree, business firms, too, are helping to improve the environment, renovate the slums, support education, science and the arts; the universities are acting to reform, and are not simply prescribing for, the ills of society; and the churches are trying to make the secular city like the heavenly one. There is not a major public or private institution in Western societies that, both voluntarily and perforce, is not broadening its conception of its appropriate responsibilities.

As the patrimonial state was validated by the authority of the past, and the liberal state by its results in the present, the technocratic state is justified by the utopian future, by the prospective improvement and eventual perfection of society. And, like the patrimonial state, it expects that the progress of society will *ipso facto* assure individual happiness, rather than the reverse relationship, as in the liberal order. For the benevolent despotism of the divinely ordained patrimonial monarch and the beneficent natural laws of the liberal society, the technocratic order substitutes the superior rationality and ethics of an authoritarian elite legitimized by its technical training and specialized knowledge. The technocrats engaged today in applying the physical and social sciences to solving the problems of society are even more certain of the absolute power and inevitable success of human reason and scientific method than the 18th-century *philosophes* and 19th-century positivists. Although the pursuit of growing and increasingly diversified goals by governments and private institutions partly annuls the functional austerity imposed by worldly asceticism, the implementation of each program is sought by more rigorous application of rational calculation and efficiency criteria than at any time since the emergence of the Protestant ethic.

Differences Among Leading Technocratic States

These characteristics of the technocratic order are manifested in varying degree among the countries of the Atlantic region owing to the differences in their historical development. The variations among the six leading nations may be briefly noted.

Despite its earlier lag at the governmental level, the United States today exemplifies the technocratic order and dramatic design to a significantly greater extent than any of the others. Its national goals are more numerous and ambitious, its governmental and private institutions more rationalized and impersonal, and its elite groups more professionalized and self-confident than elsewhere in the Atlantic region. Because of the more pervasive influence of the secularized Protestant ethic, American technocrats tend to have a stronger sense of their mission to perfect society than do their European counterparts. Moreover, as explained in the next chapter, the economic and technological achievements and the democratic pluralism of American society have fostered a more optimistic faith in the efficacy of rational calculation and a greater determination to reorganize wider and wider areas of human concern in accordance with impersonal efficiency criteria than was the case in Europe. Hence, American elites are notably more activistic, moralistic and rationalistic than European elites, and they are much more prone to utopian expectations and to prescribing scientific or common-sense panaceas.

The unique British blending in the 19th century of its originally conflicting Calvinistic and aristocratic heritages has had ambivalent effects in the 20th century. On the one hand, the combination of Puritan conscientiousness with the aristocratic norm of patrimonial service has helped to produce one of the least corruptible and most dedicated bureaucracies in the world. On the other hand, the aristocratic prejudice in favor of the omnicompetent classically educated gentleman and the empirical and pragmatic emphases of British liberalism delayed for much longer the professionalization of the administrative and managerial elites in government, business firms and other private institutions and tended to make them less imaginative and innovative—if also less utopian and activistic—than in other Atlantic countries. Resentments generated by aristocratic insistence on class distinctions counteract the strong sense of workmanship and occupational responsibility derived from the Protestant ethic. These conflicting characteristics played significant roles in the British economic difficulties of the decades after World War II.

In contrast to the English-speaking nations, the persistence of patrimonial institutions and attitudes and the much lesser extent of liberal
influences during the 19th century preserved on the continent a strong
sense of submission to the superordinate state and to the top officials
of hierarchical public and private organizations—generally the still
active members of the oldest generation—regardless of their politics
or values. Within this common characteristic, significant national differences may still be discerned.

In the postwar decades, French technocrats have come to wield
greater administrative power than do the professionalized officials of
other Atlantic countries. As inheritors of the rationalism of Descartes
and the 18th-century *philosophes,* they are second only to Americans
in their faith in reason and science and confidence in their own ability
to apply them for the renovation of society. Without their knowledge
and skills, France under the politically weak regime of the Fourth
Republic would not have recovered so rapidly from the effects of
World War II. Nor would General de Gaulle have had the resources
necessary to sustain for as long as he did the active and independent
role he wished to play in Atlantic and world politics.

Technocratic attitudes and behavioral norms are only now beginning to predominate in Germany owing to the inhibiting effects both
of the exaggerated respect for hierarchical authority intrinsic in German society and of the unhappy experiences of the German nation in
the 20th century. These factors have been sufficient to offset the influence of Germany's late 19th-century preeminence in many of the
sciences and the fact that the rationalizing and professionalizing of its
institutions commenced much earlier than elsewhere in Europe. The
current development is being accelerated by the fading of the traumatic
passivity of the German people characteristic of the postwar decades,
and by the coming to power of younger, more dynamic policy makers
in governmental and private institutions since the political changes at
the end of the 1960s. In consequence, Germany is now beginning to
overtake France in the strength of its technocratic positivism.

Until the late 1960s, the Italian economy had one of the highest
growth rates in the Atlantic region and it was steadily modernizing.
However, the transformation of the Italian political and governmental
systems lagged far behind. This disparity reflects certain peculiarities of
Italy's historical development expressed in the differences between
north and south. Both national politics and national administration
have been dominated by political leaders and functionaries from the
traditionalist south, heavily influenced by its lingering personalistic re-

lationships, the particularism of its contending families, factions and cliques, and its concern for status honor and privilege. These characteristics are perpetuated in the factional nature of Italian politics and in the lack of professionalism of most Italian civil servants, who owe their positions more to the patronage of politicians than to technical education and competence. Inadequately supported by political leaders and the overstaffed bureaucracy, the small number of well-trained young technocrats in the central government exercise much less power than their counterparts in other European countries. In contrast, the technocratization of the large business corporations, public as well as private, and even of local and regional governments in the north proceeded rapidly during the 1950s and '60s. Indeed, northern economic dynamism has been mainly responsible for the continued survival both of the national political regime and of the national administrative system despite their evident weakness and incapacity. Whether it can continue to sustain a sufficient national unity and momentum during the 1970s is an open question.

Although not part of the Western sociocultural tradition, Japanese society is becoming increasingly technocratic in a unique adaptation of traditional organic interpersonal relationships to meet the contemporary requirements for political effectiveness, rapidly rising productivity in a complex industrialized economy, and growing influence in the international system. Japanese economic, governmental, educational and other modern-type institutions are in their *functional activities* as rationalized and efficient as any in North America and Western Europe. But, the relationships among the people comprising them are in most cases derived from the distinctive elements of the traditional family household, notably the powerful emotional identification of individuals with the organization, the importance of consensual consultation in decision making, and the consciousness of the rank order of individuals and organizations. However impersonally they apply efficiency criteria in production, marketing, research and development, and investment planning, the managers, technicians and employees— and, significantly, their families as well—of a typical large Japanese enterprise are bound together by strong ties of mutual loyalty and all-inclusive responsibility. These characteristics are expressed, for example, in the practices of lifetime employment and of payment and promotion by seniority; in the provision of comprehensive welfare benefits that usually include housing, medical care, educational assistance, retirement, recreation and vacation facilities, and marriage arrangement; and in the hostility among competing organizations and their people individually. Relationships, too, between the government and the private sector resemble those in patrimonial states. Indeed, in contemporary Japan, the subordination of private economic organizations to the state and the former's converse ability to use governmental power to advance their interests are as traditionalistic as in Europe before the 19th century. Moreover, these public-private relationships are still largely governed by custom and have not been defined, restricted or superseded by specific legislation and systematic legal rules and regulations. Japan's exceptional and obviously highly effective blending of patrimonial and technocratic characteristics not only reflects the indigenous sociocultural factors that made it the first non-Western country to modernize successfully but also helps importantly to shape its future development and participation in the international system.[23]

The Technocratic State and the New Nationalism

Technocratic ways of thinking and acting are continuously engendered in the elite groups by their family experiences and education in the physical and social sciences and new administrative and decision-making techniques, and by the pressures and constraints that shape the functions they perform in the rationalized and impersonal organizations of their societies. Although, as explained in the next section, some elite-group members react negatively to these influences, technocratic attitudes tend in the majority of elites increasingly to predominate over older religious or aristocratic values and norms and the doubts recurrently generated by romantic or humanistic protests.

Technocratic values and expectations are inculcated among the general population not only by similar family, educational and work experiences but also by the opinion-molding role of the elites and the dramatic changes of the 20th century in the socioeconomic conditions and political importance of the great mass of the people. At the same time, however, their rising consumption and living standards combine with the increasingly routinized and mechanistic nature of their occupational activities in factories and offices to weaken the hold of the gospel of work and lessen the willingness to perform job responsibilities conscientiously. More and more, people want the fruits of redemptive activism without paying the price of observing the rationalized behavioral norms that have been developed in the technocratic order from the worldly asceticism of the Protestant ethic.

These parallel and ambivalent trends among the elite groups and the people generally interact in complex ways to lead to the continuous expansion of the national goals and social values that Atlantic nations are now striving to realize. The aims of national policies in the current period may be grouped into six categories:

1. *maintenance of economic growth and full employment,* requiring adequate public and private capital formation, training of labor, and research and development for technological innovation.

2. *continuing improvement of living standards,* including not only the traditional component of rising material consumption but now also better health, improved and continuous education, greater leisure and more facilities for recreational activities, and earlier and more secure and satisfying retirement.

3. *more equitable distribution of income,* particularly to eliminate poverty among the lowest income groups and to revitalize depressed and stagnant districts and towns within the country.

4. *conservation of natural and man-made environments,* including urban renewal and improvement, reduction and eventual elimination of air and water pollution, etc.

5. *advancement of knowledge,* both basic and applied, especially medical research, the exploration of space and of the oceans, and the development of additional synthetic materials and new sources of energy.

6. *safeguarding national security and meeting international responsibilities,* including defense expenditures, military research, arms assistance to allies and friends, foreign-policy subsidies, development aid, etc.

The importance of these national goals varies considerably among Atlantic countries. Owing to its more technocratic character and its status as a superpower, the United States seeks all of them and in greatest degree. In other Atlantic countries, some are still of comparatively minor significance, for example, 4, 5 and 6 in the smaller, less industrialized nations with lower incomes.

While the commitment to these national goals and the conviction that they can be achieved are general sociocultural phenomena, they manifest themselves as specific problems in the economic and political systems of a national society. The reason is that realization of these goals involves claims on resources, which are allocated both through the market and by political decisions. *This is why economic growth is of such central importance in contemporary Atlantic nations.* It is simultaneously the source of additional resources for achieving national aims, a goal of national policy in its own right (since it competes for resources with other objectives as indicated in 1 above), and an important contributor to domestic and regional economic problems (discussed in Chapter VI).

In market economies, much of the competition for scarce resources is resolved by the complex interactions among buying and selling and saving and investing activities as they are carried on by employing and employed organizations and individuals. However, the basic changes in institutions and values in the course of the 20th century have infused the transactions of the market with pressures and incentives, constraints and regulations that reflect explicit and implicit national decisions regarding the purposes for and the amounts and ways in which resources are to be used by the private sector. Equally important, the activities of governments—the public sector—have grown enormously and directly constitute major claimants for resource allocations. These two developments have brought about much greater *politicizing of the process of resource allocation* than existed in the past. In more and more Atlantic countries during recent decades, conflicts over national goals, the priorities among them, and the means for achieving them have become the major issues of national politics.

These developments have in turn continuously increased the *significance and power of the central institutions of the nation-state*. The government's responsibility for promoting the achievement of national goals and the widening scope of these objectives require it to be active over ever-broader areas of the society's life and to penetrate ever more deeply into the inner workings of the other institutional systems. Moreover, the imperative need for effective *macro*-economic management to cope with the problems generated by efforts to realize national goals—as well as by other contemporary difficulties, domestic and foreign—reinforce this trend toward strengthening the importance of the decisions and activities of the central authorities in all Atlantic nations.

These three interrelated trends in the current period—the expanding size and diversity of the national goals to be realized as quickly as possible, the increasing politicizing of the process of resource allocation, and the growing importance of the agencies and activities of the national government—are the main sources of the *new nationalism* characteristic of the Atlantic countries in the present period. This phenomenon is designated as "new" to distinguish it from the kind of nationalism that prevailed in the 19th and early 20th centuries. In contrast to the latter, which in greater or lesser degree was aggressive, expansionist and often xenophobic, the contemporary form tends to be defensive and, in most cases, without hostile feelings toward the other nationalities in the Atlantic region.

The new nationalism is expressed today in various ways. As the productivity of the Atlantic economies grew by 50 percent during the 1960s, popular expectations of early realization of social values and national goals were accordingly stimulated, and impatience of difficulties and delays was correspondingly increased. The objectives of greatest importance to the majority of the people—higher living standards, more equitable distribution of income, improvements in the natural and man-made environments—involve changes inside countries, and their external implications are neither readily nor immediately apparent. Hence, popular attention and concern tend to be directed inward rather than outward. Moreover, they are increasingly focused on the domestic political process, which helps to determine the allocation of resources among competing goals, and upon the institutions of the national administration, which are the principal agencies for carrying out such decisions. The result is more diversified and intensive—if not always numerically greater—popular participation in the political process, in the broadest sense of the term, and the resort to direct and unconventional—sometimes illegal—means of applying pressure (e.g., strikes, protest demonstrations and marches, riots, etc.), especially by the groups that are most dissatisfied or that feel most strongly threatened by the changes demanded by others.

For their part, the younger, more technocratic elites are fully aware that fulfillment of their sense of vocational mission to perfect society requires them to direct both the process of resource allocation and the agencies of government that implement its determinations. In consequence, their attention and concern, too, are drawn inward—by the disputes over goals and resources, by the domestic conflicts and difficulties impeding social progress, and by the need to develop and administer the measures for overcoming them.

But, especially in the United States, the redemptive activism of the technocratic elites tends to have a wider frame of reference as well. Their education and training and their occupational responsibilities inculcate in them both the perception of, and the motivation to protect and advance, the interests of the organizations in which they work and of the nation as a whole, as they conceive them. These interests relate not only to domestic developments but also to external relationships in the regional and worldwide systems. Moreover, elite-group perceptions and conceptions of private and national interests are always infused, distorted and magnified by the elements comprising the dramatic design of their societies. Their sense of redemptive mission always in some degree reinforces the calculations of rational interest in impelling their countries to active efforts to influence developments inside other nations, as well as the latter's behavior in the regional or worldwide systems. The self-confidence born of positivistic faith in the power of reason and science always in some way obscures their perception of the realities of the external situations with which they are trying to cope and in some measure misshapes the means they believe effective for doing so. How active and independent a role Atlantic countries seek to play in the regional and worldwide systems, and how its form and content are molded by the interaction of specific rational interests and the sense of national identity and purpose, also depend at any particular time upon the nature and relative urgency of domestic pressures and opportunities compared with those in the external environment.

The Prospects for the Emerging Technocratic Society

Before turning to these subjects in the subsequent chapters, an obvious question regarding the current transformation of Western societies needs to be answered. Can the technocratic state continue to prevail when, even before it has reached its fullest development, it is already under serious attack in many Atlantic nations, especially the United States? I believe that it can and will because, with certain exceptions discussed below, the attack on the technocratic order—like that order itself—is an expression of redemptive activism, and the alternatives pressed by dissident groups require use of the same means as are intrinsic to technocratic positivism. Thus, in important—although usually unintentional—ways, even the enemies of the technocratic order are hastening its predominance.

At bottom, this paradox is a contemporary manifestation of Western civilization's fertile ambivalence of trends and countertrends that has been so largely responsible for its self-instigating dynamism and unique development since the 11th century. Moving rapidly today toward realization of goals hitherto unattainable, Western society is characteristically questioning whether the effort is worthwhile. Convinced as never before that reason and science alone provide certain knowledge for mastering nature and perfecting society, Westerners are also searching for meaning in the mystical, the magical, the instinctual, and the Oriental, as well as in formalistic and pentecostal religions. Confident that rational calculation and rational compromises among competing interests constitute the only effective approach to the solution of social problems, the Atlantic nations are plagued by outbreaks of violence, irreconcilable extremisms and self-destructiveness. With their optimistic faith in the inevitability of progress validated by unprecedented material and intellectual accomplishments, members of elite groups suffer from pessimistic forebodings of social entropy and feelings of personal alienation. Imbued with a righteous sense of its redemptive mission, Western society is shaken by moral doubts and the anxiety of guilt. Addicted to self-congratulation, it is deeply critical of its own failings.

These contradictions and perplexities are reflected in, and are in turn stimulated by, conflicts over particular goals, institutions and behavioral norms. The great majority of these disputes are explicitly or implicitly premised on acceptance of the social order and values of the technocratic state. Whether they recognize it or not, the participants are contending not over whether to preserve or abolish the technocratic order but over the priorities among its competing national and group objectives, the relative magnitudes of the resources allocated to them, and the comparative effectiveness and moral worth of the means chosen for achieving them.

Many of the most conspicuous and important conflicts today are wholly or predominantly of this kind—the struggle against racial and other kinds of discrimination, the drive to overcome poverty and the effects of cultural deprivation, the demand for rising mass consumption and greater leisure, the campaigns to improve education and to protect the natural environment, the pressures to rebuild the cities, as well as the resistance of particular groups and localities to such changes. Those who seek greater resources for one or several of these goals usually demand that they be denied to or diverted from the others, and they generally insist that certain objectives—notably national defense and military operations abroad, foreign-policy subsidies, development aid, the exploration of space, etc.—be given the lowest priorities, or none at all. Conversely, those to whom the latter activities are of paramount importance fight to preserve or to increase the shares of available resources allocated to them. Such disputes are typical manifestations of the politicizing of the process of resource allocation in the period of the new nationalism.

Controversies over the means by which the technocratic state tries to achieve competing objectives are in most cases also based, consciously or unconsciously, on continuation of the essential features of its institutional system and behavioral norms. The rationalized and impersonal organizations, the insistence on rational calculation and universal application of efficiency criteria, the new computer-based decision-making techniques and information systems—these and other institutional and cultural characteristics of the developing technocratic society are denounced as dehumanizing, stereotyping, faceless, soulless, mechanistic, pecuniary. Such judgments do indicate a more serious dissatisfaction with the technocratic order than the disagreements over resource allocations to competing goals. Nonetheless, these critics are impelled by their own sense of redemptive mission to reform the evils they condemn, and they press their efforts with the same logical rigor and with equally utopian expectations of early and easy success as their opponents.

Moreover, their remedies usually require use of the very organizations and methods they decry. With the exceptions discussed below, there is no presently sought national goal or social value that does not generate a significant claim on resources—the advancement of art and science as well as the construction of popular entertainment facilities and highways, greater leisure and better health and education as well as the increase of material consumption, public investment in urban improvements no less than private investment in new factories. Because competing demands are more pervasive, powerful and impatiently pressed than ever before, they cannot be relieved for long simply by redistributing resources but only by increasing the total available. Thus, however much they may condemn it, economic growth is as important to the critics of the technocratic order as to its defenders. The pressure for growth comes as much from those who denigrate it as "the GNP rat race" and "a false, materialistic standard" while insisting on greater resource allocations to enhance "the quality of life" as from those who recognize that it is an essential means for mitigating social conflicts and realizing social goals.

The crucial importance of an immense and growing volume of resources in turn makes much of the institutional structure and many of the behavioral norms of the technocratic order similarly indispensable. High productivity, mass production and rapid technological innovation are inseparable from diversification and the division of labor, mechanization and automation, large rationalized and impersonal organizations, the use of rational calculation and efficiency criteria in decision making, mass markets, expensive research and development, and adequate incentives for entrepreneurial vigor and conscientious work performance. Critics of "dehumanizing" mass production, of the "consumer society," of "organization men" with their "computer mentalities," usually fail to draw the full implications for the achievement of their own welfare goals of changing or abolishing the institutional and behavioral characteristics they denounce.

The opponents of the technocratic order who do recognize the price that would have to be paid for its radical transformation are a comparatively small minority. Even among them, a distinction has to be made between those who condemn it root as well as branch and those who are intent only upon replacing one form of technocratic society by another. In the latter category are the Moscow-oriented Communist Party members and sympathizers; and the Maoist, Trotskyist and other Marxist factions in the New Left, who look to China or Cuba as paradigms—or to some hypothetical model of a socialist order, in which the replacement of large private organizations by large government bureaucracies is supposed somehow to make possible the maintenance of high productivity and rising mass consumption without sacrificing humanistic values.

In contrast, the truly radical dissenters are prepared to reduce both population and living standards as the precondition for abolishing the hated technocratic order. For its high living and plain thinking, they expect to substitute some form of plain living and high thinking. This motivation is characteristic of the young people and others seeking interpersonal satisfactions in the emotional warmth and altruistic sharing of small philadelphic communities; and of the neoanarchists, who anticipate the harmless fulfillment of diverse individual potentialities in a new Eden devoid of psychological repression and social constraint. However, other radical dissenters are simply negative, like the nihilists bent upon destroying and not replacing the technocratic state. Finally, there are the dropouts—those who cease to cope positively or negatively with the technocratic order and withdraw, figuratively or literally, to its interstices through solitary living, drugs or psychosis.

The pervasive power and momentum of institutions, self-conceptions, values and behavioral norms as deeply rooted historically and psychosocially as those of the technocratic order are exceedingly great. They are continuously strengthened not only by the activities of the majority, who accept them, but also by the efforts of most of their ostensible opponents, whose own senses of mission help to sustain expectations of social perfection and personal happiness and to intensify the pressures for increasing resources and for more rationalized organizations and decision-making methods. Hence, in one mode or the other, redemptive activism and positivistic ways of thinking and behaving will continue to be overwhelmingly predominant in most areas of contemporary life: in the natural and social sciences and their applied technologies, in government and politics, in business and other private economic activities, in the mass media and popular entertainment, in the educational system—even in philosophy, if not nearly to the same extent in art, literature and religion.[24] Thanks to the influence of the critics of the technocratic order, its fuller development in the years ahead is likely to be shaped by greater regard for humanistic values and growing concern to preserve—perhaps even to improve— the humaneness of institutions and relationships. These effects should become increasingly evident as the age groups educated after World War II, and hence more deeply imbued by their life experiences and formal training with the sense of mission and the requisite technocratic skills, reach the top levels of governmental and private institutions in the late 1970s and the 1980s.

Thus, the various fusions and developments since the 16th century of the Hebraic injunction to work for the perfected earthly kingdom and the Hellenic insistence on human reason as the means for understanding nature and society have played crucial roles in the unprecedented intellectual, social and technical achievements of Western civilization. But, they have also contributed to its failures, existential and moral. They continually generate utopian expectations and addiction to panaceas, which inevitably result not simply in waste and delay but, more important, in greater frustration and suffering than would have been involved in more reality-oriented approaches. The arrogance and self-righteousness which they engender have led to abuses of power, intolerance, repression and callous disregard or brutal exploitation of human life and individual integrity—all in the names of reason, science, progress and perfectibility. Although the harm they have done does not match that perpetrated by religious and racial fanaticisms, by imperial ambitions, by revolutionary and counterrevolutionary retributions, and by totalitarian orthodoxies, nevertheless the potentiality for evil of self-confident positivism and redemptive activism should not be underestimated, nor can its manifestations be excused by the good intentions that often motivate them.

III

Dramatic Design and Foreign Policy in the Soviet Union and the United States

NATIONS THAT HAVE PLAYED leading roles in world politics have been impelled to do so not simply by their perceptions of national interest but also by the self-conceptions and motivations comprised in the dramatic designs of their cultures. And, because cultures differ, so too do national senses of identity and of purpose and the intensities with which they are expressed. In the decades since World War II, the competing efforts of the two superpowers, the Soviet Union and the United States, to satisfy their interests and express their redemptive missions have been of paramount importance in the international system. This chapter traces the evolution of their dramatic designs and analyzes the changing ways in which these self-conceptions have been affecting their foreign policies and external behavior and are likely to do so in the future.

Soviet and American policies and actions are important for two reasons. First, in different ways and degrees, each superpower will go on exercising the major influence within its regional system even though the changes of the 1960s have lessened its preponderance and increased the desire and capacity of its associated states for greater freedom of action. Second, notwithstanding these changes, prospects for world peace and war will continue during the 1970s, if not thereafter, to be affected by the behavior of the Soviet Union and the United States to a more significant extent than by that of other participants in the international system.

The Sociocultural Sources of Soviet Messianism

Russian perceptions and conceptions of the nature of the international system and of the role that the Soviet Union has to play in it have twin roots in the history of Russian society and culture and in the Leninist adaptation of Marxism.

Despite many centuries of subsequent development, Russian institutions and ways of thinking still reveal much of their original inheritance from one of history's most durable patrimonial states—the medieval Byzantine social order and dramatic design. The Byzantine state was characterized by caesaropapism—that is, it was headed by an absolute emperor who was God's vicegerent on earth to rule over both the secular society and the church—and it was supported by an elaborate hierarchy of court-centered officials. This Byzantine patrimonial model was adapted by the Muscovite rulers to meet their own needs during the crucial three-way struggle in which they were engaged from the 15th through the 17th centuries: to free themselves from Tartar suzerainty, to conquer the other Russian principalities, and to assert their absolute authority over the boyars, the local landed magnates.

In accomplishing these three objectives, the Muscovite patrimonial regime was also aided by the developing Russian sense of national mission derived from the messianic expectations of Byzantine Christianity. These ideas were crystallized in the concept of Moscow as the "Third Rome." Because the first holy imperial city, Rome, was in the hands of the Papal schismatics, and Constantinople, the second holy imperial city, had fallen to the infidel Turks in 1453, Moscow became the third holy imperial city designated by God to lead all mankind into the true Orthodox faith. With the national consciousness centered in the quasi-divine figure of the Tsar, "the Autocrat of all the Russias," Berdyaev explains:

> The Russian religious vocation . . . [was] linked with the power and transcendent majesty of the Russian State, with a distinctive significance and importance attached to the Russian Tsar. . . . The Third Rome presented itself to [Russian] minds as a manifestation of sovereign power, as the might of the State. . . . There enters into the messianic consciousness the alluring temptation of imperialism.[1]

These ideas have provided part of the motivation and much of the moral justification for Russian imperial expansion in Asia and Europe since the 17th century. They also reflected and in turn helped to strengthen the relationships of authority and subordination in Russian society evolved under Ivan the Terrible and Peter the Great and further developed during the ensuing two centuries. Their main features were the absolute authority of the patrimonial ruler, the passive acquiescence of the great mass of the people, and the replacement of the autonomous local nobility by an expanding official bureaucracy rigidly bound by policies and procedures emanating from the Tsar and his immediate advisors. The system reached its most autocratic form in the mid-19th century under the personal rule of the "Iron Tsar" Nicholas I.

Although Russian messianism justified and encouraged imperial expansion externally and Tsarist absolutism and aggrandizement of the patrimonial state internally, its predominant expression did not envisage a millennial kingdom of this world but rather an Augustinian redemption beyond space and time. This transcendental expectation mainly reflected the painful tragedies of Russia's historical experience.[2] In part, too, it was fostered by Russia's deeply ambivalent attitude toward the West.

Knowledge of West European social and cultural achievements and the desire to imitate them began to be significant factors in Russian development during the 17th century. Emulation of the West was openly manifested in the reforms of Peter the Great in the late 17th century and was greatly stimulated by the spread of Enlightenment ideas in the 18th century. Western influence was further strengthened by the modernizing efforts of the two Alexanders and the start of industrialization in the 19th century. It culminated in the dynamic impact of romanticism, socialism, anarchism and other dissident Western ideas on the diverse circles of liberal and radical writers, publicists, artists, political and social thinkers, and academic and professional groups whom the Russians call collectively "the Intelligentsia."

The ideas and achievements of the West constituted, on the one hand, a model to be admired and copied and, on the other, a constantly painful reminder of Russian backwardness and cultural dependence. The West was at the same time the source of new and better ways of thinking and living and of alien ideas and institutions capable of destroying the traditionalistic agrarian society on which depended both Russia's sense of unique identity and the stability of its patrimonial order. Western Europe was simultaneously idolized and feared—imitated for its manifold achievements and rejected for its heresy, materialism and degeneracy. On the one hand, Russia was to be despised for its ignorance, poverty and scientific, economic and political deficiencies; on the other, it was to be exalted for its doctrinal purity, its spiritual superiority, and its altruistic messianic mission. If Russia's material, technical and socioeconomic accomplishments were negligible compared with those of the West, it was because the latter resulted from the competitive individualism and self-seeking of the West Europeans, while the genius of the Russian people was expressed in their religious and spiritual concerns and in their sense of collective identity and communal fulfillment.

These complex and powerful feelings of inferiority and superiority and attraction and repulsion *vis-à-vis* the West played a major role in the development of Russian attitudes and policies during the 19th and 20th centuries. With appropriate changes in terminology, they have been carried over from the official Tsarist explanation of national identity and purpose to that of the Soviets. In different forms and varying degrees, they have also been expressed in dissident movements among the Intelligentsia, such as populism, pan-Slavism and socialism, before the Bolshevik Revolution, and in the humanistic and positivistic stirrings against Soviet absolutism and conservatism in the years since Stalin's death.

The orthodox Soviet version of national identity and purpose is an adaptation of Marxism to the distinctive characteristics of Russian society and culture. Under an imported terminology and new institutional arrangements, it has perpetuated both the historical forms of Russian social relationships and the traditional conceptions of Russian identity and messianic mission.

Although there were beginnings of liberalization under the two Alexanders and after the Revolution of 1905, Russia did not have anywhere near the degree of movement toward a liberal order experienced by even the German and Austrian empires before 1914. The Russian patrimonial order was still largely intact when the Bolsheviks seized political power in 1917 and, in all essentials, the Soviet regime inherited the centralized autocracy of the Tsars.

The Soviets soon replaced the institutions of the Tsarist patrimonial state by new organizational arrangements better able to carry out the total mobilization of Russian society for accelerated industrialization and to assure the autocratic rule of the communist elites. This Soviet totalitarian development of the patrimonial state is, however, less complete, even though much more effective, than the Tsarist regime. It is less complete because Russian society is today more differentiated institutionally—although not in terms of social classes—as a result of industrialization and urbanization than was the homogeneous agrarian society of the past. It is more effective in consequence of the greater efficiency of modern techniques and instruments of communication and social control and the superior training and morale of the ruling communist elites compared with those of the Tsarist bureaucracy. Yet, virtually all policy making and innovation in the economic, political and other institutional systems of the society still flow downward from the top of the ruling hierarchy. Thought and expression in artistic, literary and scientific endeavors continue to be controlled to ensure conformity with official doctrines. Even though over half a century of Soviet rule has transformed Russia organizationally and economically, the basic relationships of authority and subordination, the distribution of political power, and the norms of behavior reflecting and supporting them continue substantially as they were before the Revolution.

Nor do the much more extensive urbanization and spread of education under the Soviets appear to have generated significant pressure among the great mass of the Russian people—as distinct from certain sections of the Intelligentsia and the new technocrats—for decentralization of decision making, or even much more willingness on their part to take initiative and to participate in the various levels and institutions of policy making. Passionate mystical devotion to country—to the endless, open, fertile, mysterious land, to "Holy Mother Russia so harsh and so kind to her children"—has always been deeply ingrained in the Russian people. But, as in most patrimonial states, their attitude toward their successive political rulers has usually been one of passive acquiescence rather than of strong positive loyalty. Only in periods of great national peril—as during the Napoleonic and Nazi invasions—has love of country transformed resignation toward the regime into active support. Today, both mass support for and mass resistance to the Soviet autocracy seem to be as passive as they have generally been throughout Russian history.[3] Active protest against totalitarianism, begun under the Tsars in the 19th century, continues to be carried on mainly by limited groups of the Intelligentsia, with great courage and considerable ingenuity but so far with little lasting effect on the attitudes and policies of the ruling Soviet elites.

Marxism has, however, made one very significant change in Russia's sense of mission. While reinvigorating traditional Russian messianism, Marxism also secularized it and focused Russian redemptive activism on progress and perfectibility in this world. At the same time, it condemned and endeavored to suppress the other-worldly expectations of Russian Orthodox Christianity, as well as the nostalgic yearnings for idealized peasant communes and mystical philadelphic communities of the 19th-century populists, anarchists and socialists.

Nonetheless, despite the great economic growth and substantial improvement in living standards and the very considerable scientific advances under the Soviet regime, ambivalent feelings toward the West persist, especially among the communist elites. Indeed, these conflicting attitudes have been intensified in Leninism and Stalinism. For, the West—since World War II, primarily the United States—is not simply the source of heretical and subversive ideas, as it was in past centuries. It is now also an enemy believed to be actively conspiring against the Soviet Union and ceaselessly trying to block fulfillment of Russia's world-transforming mission. Western achievements still evoke both emulation and envy, giving rise to feelings of inferiority and to exaggerated or fictitious claims of Russian scientific priority and Soviet social and moral superiority.

A comforting theory in Western Europe and the United States maintains that economic growth and social evolution will inevitably erode Soviet totalitarianism, weakening the conceptual and motivational grip of Russian Marxism, and resulting in an increasingly pluralistic society that will follow more humane and rational policies at home and abroad. Certainly, as explained in Chapter II, an economic system grows by becoming more differentiated and complex through innovation and increasing rationalization of its organizational forms and methods of operation. This developmental process means that interests become more diversified and decisionmaking more decentralized and dispersed, and that both become more efficient. In other words, the more industrialized, urbanized and educated a society is and the more differentiated, interdependent and rationalized its constituent social units and institutions are, the more it will manifest the liberalizing characteristics of pluralism.[4] This process has been operating in the Soviet Union during recent decades, and it lies at the root of such liberalization as has actually occurred. By its nature, the consequences of this process are most evident in the economic system where, in recent years, a start has been made in delegating somewhat greater scope for decision making to individual enterprises and in measuring their performance by more significant criteria than physical output. Grudgingly conceded by the Soviet hierarchy, these changes have resulted from the manifest inefficiency of detailed central planning in an increasingly complex economy and from the positivistic ways of thinking of the younger and more technically trained managers and economists.[5]

However, the effects of this process are continually being inhibited and offset or deliberately nullified by the momentum of patrimonial relationships of authority and subordination and their related behavioral norms as they are perpetuated in contemporary Soviet institutions and Marxist conceptions and motivations. It is essential to recognize that both the ruling elites and the Russian people generally have been habituated by the experience of five centuries to centralized autocratic power, mass passivity, and a unanimist climate of opinion. In these circumstances, neither rulers nor people can rapidly learn to tolerate, much less to desire, the uncertainties and risks of popularly responsive political institutions, of self-instigating and self-responsible individual and group initiatives dispersed throughout the society, and of freedom to express diverse attitudes and divergent interests. Such manifestations of pluralism are continually being stimulated by the developmental process. But, periodically, the feelings of insecurity and anxiety they provoke at all social levels become strong enough to lead to repressive measures by the ruling elites, in which the great majority of the Russian people acquiesce. Repression of liberalizing tendencies reflects not only the fear of a threat to the communist regime but also a deeper sense that heterogeneity and individual autonomy are immoral, contrary to the orderliness and seriousness of purpose of a well-behaved communal society.

Thus, the two opposing tendencies in Russian society are almost in equilibrium and, in consequence, the developmental process is likely to take much longer to transform Soviet totalitarianism than, in the course of the 1960s, it became fashionable in the West to suppose. Periods of liberalization, as under Khrushchev, will continue to be followed by renewed repression, as under his successors. A long-term trend toward gradual amelioration may perhaps be discerned in the fact that the post-Khrushchev repression was not as extreme as the terrorism of the 1930s and of Stalin's last psychotic years. Nevertheless, the pace of the long-term liberalizing trend is very slow and its cumulative effects are generational in their manifestation.

In this connection, it is significant that many of the leading proponents today of the "hard line" are younger Communist Party functionaries born after the Revolution and now rising to top positions in the hierarchy. Their conservative attitude is a product of the values, behavioral norms and fears inculcated during their childhood and adolescence. And, it has been reinforced by the hazardous selection process —compounded of rigorous characterological and doctrinal tests and of Byzantine palace intrigues and factional conspiracies—through which they have reached the upper levels of the Party. Unlike the older generation, many of whom spent years of exile in Western Europe and were often well-read in Western political and social theorists and philosophers, the upcoming age cohorts of future Party leaders have little, if any, personal experience of the West and know only a carefully censored selection from its dangerous writings. Their parochialism constitutes another factor inhibiting liberalization and helping to perpetuate the existing blend of traditional Russian and Marxian perceptions and conceptions.

True, positivistic ways of thinking, more reality-oriented than Marxism, are increasingly fostered by the technical and scientific education of the younger Party functionaries—as well as of the more numerous non-Party managers, technicians and scientists—and by the nature of the problems with which they have to cope in a more and more complex and interdependent economy. However, although positivistic influences should eventually help to make Russian perceptions and conceptions more realistic, they are not necessarily liberalizing in other respects because positivism, as predominantly an elitist way of thinking, is inherently authoritarian and tends toward messianic—as well as simply reformist—activism.

Soviet Foreign Policy and World-Transforming Mission in the Postwar Period

In its influence on Soviet foreign policy, the Russian version of Marxism has perpetuated and strengthened the effects of traditional Russian attitudes, aspirations and fears. By secularizing Russian messianism and reorienting it toward achievement in this world, Leninism substituted for the transcendental religious validation of Russian imperialism a justification based on the dialectic of history. Faith in the inevitability of final victory and the determinative role assigned by dialectical materialism to the inexorable workings of the historical process provide a rationalistic reinforcement for traditional Russian patience and a plausible rationalization for the failures and frustrations of Soviet foreign policy. Soviet setbacks need not be, and have not been, attributed to the deficiencies of Leninist and Stalinist doctrines. Instead, they are believed to result either from the evil machinations of the capitalist enemy, who has seduced the responsible officials into "counter-revolutionary" or "reformist" ways of thinking, or from unfavorable "objective conditions" (e.g., the absence of a "revolutionary situation").

The consequences for Soviet foreign policy of this conceptual framework have been ambivalent. On the one hand, Soviet analysis typically has a long-term perspective and seeks to take into account a broad range of political, social, economic and "ideological" factors. The ready availability of satisfying explanations gives Soviet policy makers considerable *tactical* flexibility in starting and terminating initiatives, reversing alliances, and abandoning unfavorable positions. On the other hand, Marxist categories of analysis are seriously deficient in empirical validity and forecasting accuracy, and the insistence on Leninist or Stalinist orthodoxy makes Soviet foreign policy *strategically* inflexible, envisaging eventual achievement of the long-range objective of establishing Soviet-dominated communist regimes throughout the world.

Although the commitment to this world-transforming goal goes back to Lenin's founding of the Third International after World War I, significant progress toward its realization became possible only in the situation existing in the aftermath of World War II. The years from 1945 until the early 1960s were two decades of basic reconstruction of an international political and economic system which had been largely shattered by the great depression of the 1930s, the interwar aggressions of Italy, Japan and Germany, and the immense destruction and disruption of World War II. During the postwar years, too, there was not only a nearly universal conviction that a new world order—more peaceful, just and prosperous than the old—had to be constituted. There were also two competing designs for such a new world order—one explicit in the expectations and policies of the Soviet Union, and the other implicit in those of the United States as explained later in this chapter— and each was backed by a powerful nation possessing the will to try to realize its conception.

To the Soviet Union, World War II was the inevitable outcome of the deepening crisis of the capitalist system, which had been driven to war by its inability to overcome the mass unemployment of the great depression and by its failure to prevent the aggressions of German, Italian and Japanese fascism. The Soviets believed that the bankruptcy of capitalism and its further debilitation by the war opened the way for proletarian revolutions in the capitalist countries and for revolutions of national liberation in the colonial regions of Asia and Africa and the semicolonial states of Latin America. Regarding itself as the "socialist motherland" and the organizing and directing center of the world communist movement, the Soviet Union provided the leadership and the resource base for these two types of revolution against the existing international system of capitalist imperialism. It guided and assisted the local communist parties and extended considerable credits to the governments of newly independent Asian and African nations—as well as to Cuba after Castro's seizure of power—whose foreign policies and domestic developments it expected thereby to control.

In the Soviet view, although capitalist imperialism was suffering from self-inflicted and ultimately fatal wounds, it was nevertheless still capable of organizing a most dangerous attack upon the Soviet Union—indeed, its "inner contradictions" would sooner or later compel it to do so. Hence, in its own national interest as well as to advance the revolution, the Soviet Union was convinced that it had to push the frontiers of communism as far westward in Europe as possible, prevent the resurgence of Germany and Japan (the former "spearheads of capitalism" against the socialist motherland), undermine the efforts of the United States to form anti-Soviet alliances in Europe and elsewhere, and support the liberation movements in Asian and African countries, which would inevitably gravitate into and strengthen the socialist camp after attaining their independence. The Soviets anticipated that the United States, as by far the wealthiest and strongest capitalist power, would be impelled to provide leadership and resources for resisting communist advances and organizing anti-Soviet movements. Thus, the United States was regarded as the main enemy endangering the Soviet Union and blocking the construction of a new, peaceful and progressing socialist world order.

How actively and by what means the Soviet Union seeks at any given time to fulfill its messianic mission are tactical, rather than strategic, considerations. Hence, in contrast to the rigidity of their long-term expectations, Soviet policy makers have been quite flexible in adapting their tactics to changing internal and external constraints and opportunities. The next chapter analyzes those developments in the international system as a whole which helped to bring the postwar period of reconstruction and cold war to an end by the mid-1960s. Here, the internal trends in the Soviet Union that also contributed to moderating its world-transforming efforts may be briefly described.

Like the nations of North America and Western Europe discussed in Chapter II, the Soviet Union was confronted with increasingly difficult problems of resource allocation in the course of the 1960s, although they took different forms from those of the West. For the Soviet Union, the achievement and maintenance of nuclear parity with the United States, the ancillary rivalry in the exploration of space and other costly scientific and technological fields, the provision of substantial military and economic assistance to client states in Asia, Africa and Latin America, and other aspects of its foreign policy and external behavior necessitated increasing claims on available resources. In addition, Soviet citizens expected improvements in living standards after the long years of enforced austerity required to make possible the very high rate of capital investment and the reconstruction of war devastation. At the same time, however, the Soviet Union was unable to achieve, despite massive investment, a rate of economic growth adequate to meet all of these rising claims on resources. Aggravated by the poor harvests of the mid-1960s, the lag in the growth rate was—and continues to be— especially serious in agriculture. But even in industry, various factors combined to raise the incremental capital/output ratio and to inhibit increases in the productiveness of labor. Some are technical economic factors relating to the extent and composition of Soviet industrializa- tion, the relative obsolescence of much of the machinery and equipment in many branches of industry, and the slowness of technological appli- cation despite the advances of Soviet science. They are also in part reflections of the more basic institutional limitations of an economic system dependent upon detailed central planning and control and of a culture not conducive to individual initiative and self-responsibility.

Thus, after 1960, the Soviet regime became concerned about in- ternal economic problems more difficult to deal with than those of reconstruction in the postwar recovery period. Both the expectations for rising consumption and the more positivistic ways of thinking of the technically trained elites have continued to generate pressures for re- solving these perplexities in ways that are more responsive to domestic needs and rely upon more rational arrangements and techniques. At the same time, however, the nature and extent of the liberalization per- mitted by the Soviet regime have been insufficient to relieve either the pressure on resources or the pressure for reform.

These internal problems reinforced the effects of developments in the international system during the 1960s in muting the efforts of the Soviet Union to fulfill its messianic mission. But, continued devotion to the long-range world-transforming goal is rooted in the nature of Russia's totalitarian order and in the dramatic design of its culture. The liberalizing tendencies generated by industrialization, urbanization, education and the increasing positivism of the managerial and technical elites are likely to be insufficient to counteract decisively the momentum of patrimonial relations of authority and subordination and of mes- sianic ways of thinking and acting for at least another generation— perhaps not until well into the next century. Hence, Soviet policy makers will pursue their world-transforming goal more openly and intensively whenever they perceive the tactical situation, internal and external, as propitious for doing so. The likelihood of such behavior is evidenced by the scale and importance of Soviet involvement in Egypt, India, Vietnam and other countries in recent years, and by the persisting effort to develop "all-ocean" naval capabilities. There will undoubtedly be other expressions of the Russian sense of redemptive mission in the years to come.

The Sociocultural Roots of American Redemptive Activism

The perceptions and conceptions that play major roles in the making and execution of American foreign policy had their origins in the unique historical process through which the distinctive society and culture of the United States were formed. Where the historical development of Russia omitted the experience of a liberal order, or even of a substantial movement toward it, that of the United States never included a period of patrimonial order. Although the original settlements were founded in the 17th century during the patrimonial regime of the Stuart dynasty, they were free of detailed control and regulation by the central authorities in London to a very much greater degree than was the case in the French and Spanish colonies in the New World. This difference reflected the less complete and secure character of the English patrimonial state compared with those of France and Spain, and the unique role played by Puritans and other dissenters in the founding of some of the most influential English colonies. It also expressed the related ways of thinking and acting of the settlers themselves, who were already accustomed by developments within English society to self-instigating, self-reliant and self-responsible norms of behavior.

The result was that, from the beginning, the institutional systems—political, economic, religious, educational—of the 13 colonies were highly decentralized, with wide scope for self-government by the organizational units of which they were composed. The nearest approach to patrimonial relationships and ways of thinking was in the South, and that only after the rise of large plantations worked by slaves in the late 18th century. Moreover, it was the attempt by the central authorities in London to enforce more active control and more extensive mercantilist regulation of the colonies' political and economic systems that led to the Revolution against British rule.

The consequences for the subsequent development of American society of the virtual absence of a patrimonial order in the colonies were among the key factors that explain why, from the early 17th to the mid-20th centuries, Europeans settled in the United States, what they expected both to find and to achieve there, and how these motivations and expectations affected their conceptions of their country and of its role and responsibilities in the world.

From the earliest settlements through the post-World War II immigration, the great majority of the people who came to the United States explicitly or implicitly rejected the Old World and expected to make a "new beginning" in the New World. Except for the slaves brought from Africa and those indentured servants involuntarily transported to the colonies, there was a basic predisposition in the minds of Americans to think of the New World as better than the Old. This was true particularly with respect to those characteristics that the two main waves of immigrants found sufficiently intolerable in Europe to induce them to brave the perils of the Atlantic voyage and the hardships and uncertainties of the new beginning in America. The 17th- and 18th-century immigrants from England and Scotland faced the dangers of establishing themselves in the wilderness of the advancing frontier regions; the 19th- and 20th-century immigrants from Ireland and continental Europe were confronted with the alien and competitive environment of the already settled areas. Yet, in both cases, they preferred unknown hardships to familiar frustrations. Regardless of whether their dissatisfactions were religious, political, economic, or some combination of the three, virtually all expected the New World to provide them with the freedom and the opportunities unavailable in the Old.

The contrast between the Old World and the New and the concomitant psychological rejection of the former and emotional identification with the latter have constituted a fundamental and unique structural element in the conceptual framework of Americans since the early 17th century. This dichotomy has served to support and organize many particular themes, some persisting throughout American history and others important during certain periods. Together, they comprise the most universal and deeply rooted portion of the American cultural heritage significant for the formation of the sense of national identity and purpose. For more than three hundred years, children born in the United States have been continually subjected to these themes and self-conceptions from the beginning of their acculturation within the family, during their years of formal education, and throughout adulthood in their participation in the attitude-forming institutions of the society. The many variations and combinations of these themes have been expressed in innumerable sermons and speeches, pamphlets and books, plays and movies, newspapers and magazines, songs and pictures, cartoons and jokes, slogans and advertisements. Indeed, they are so familiar that we are rarely aware of thinking in their terms and of uttering them in our speech. But, this unconscious day-to-day use makes them all the more powerful.

From the early 17th to the early 20th centuries, the themes and self-conceptions were explicitly religious in content and validation; in fact, they were more or less secularized only in the interwar period. Until the turn of the century, not only did the successive opinion-leader groups in American society contain a high proportion of clergymen but also the great majority of lay elites were active participants in the religious life of their communities. Hence, the original religious formulations predominated into the early decades of the 20th century, and their echoes can still be heard today.

Although Biblical images and analogues of all kinds were widely prevalent in American culture until the 20th century, the themes dealing with the meaning and destiny of America were most heavily influenced by the messianic and millennial tradition sketched at the beginning of Chapter II. In contrast to the old, corrupt, exhausted and crowded societies of Europe, the uncontaminated, untapped and thinly peopled wilderness of America made it the natural locus for communities of Edenic innocence and plenty. Thus, it was not only the availability of cheap—and later free—land but also the perfectionist possibilities unique to unspoiled America that caused both the writers of imaginary utopias and the organizers of actual utopian communities, religious and secular, to locate most of them in the New World.

To many millennialists on both sides of the Atlantic, the settlement of the New World and the conversion to Christianity of its Indian inhabitants were prerequisites for the coming of "the last days" when the forces of evil under Satan would be decisively defeated and the Millennium would begin. Hence, for English Puritans no less than for Spanish Franciscans and Jesuits, the settlement of the New World was an event of cosmic significance, an apocalyptic development in the great struggle between good and evil that constituted the essential dynamic in man's striving for redemption in or from history. In New England from the very beginning and throughout the colonies after the Great Awakening of the mid-18th century, this conception fostered the conviction of American exceptionalism on the Hebraic model. Just as the children of Israel had been chosen by God for a special redemptive mission and were given a promised land—"overflowing with milk and honey"—in which to fulfill it, so divine Providence designated America, physically separated from the contamination of Europe, as the new promised land of plenty in which Americans, the new chosen people, would fulfill their redemptive mission both for their own salvation and for that of all mankind.

These themes have naturally influenced the American attitude toward Europe. In contrast to the deep ambivalence of Russian feelings about Western Europe, the predominant American attitude has been relatively consistent in consequence of the Old World/New World opposition. Unlike the Russians, Americans have never considered Europe to be socially or politically superior nor, since the mid-19th century, have they regarded it as technologically more advanced. True, there has always been a small elite minority—principally certain 19th- and early 20th-century literary and artistic groups and, more generally, the leisured wealthy families of the Eastern seaboard—convinced of European cultural superiority and scornful of American crudity and materialism. Nonetheless, until World War II, the great majority of opinion leaders and of the people regarded Europe with more or less suspicion and disapproval. Until then, Americans generally felt threatened by and continually warned one another against a presumed proneness on Europe's part to advance its interests at the expense of America's—a capability attributed, however, not to Europe's superior intelligence or technical proficiency but pejoratively to its Machiavellian deceitfulness, expedient morality, authoritarian political institutions, aristocratic snobbishness, greed and hypocrisy. The simple, honest, natural American may initially be at a disadvantage in dealing with sophisticated and artificial Europeans, but Yankee democratic virtues could in the end be counted on to outwit European vices. This attitude receded only when the United States emerged from World War II as the premier superpower, and the political and economic weaknesses of the West European states were clearly evident. That it has not wholly disappeared may be seen in the revival of anti-Europeanism during the U.S. trade and monetary difficulties of the late 1960s and early '70s.

The moral condemnation of Europe was the obverse of the American conviction of moral superiority that, from the very beginning of Puritan New England, has continued to constitute the first of the three main elements comprised in the American sense of mission. On the ship bringing him across the Atlantic in 1630, John Winthrop, the first Governor of the Massachusetts Bay Colony, wrote:

> Men shall say of succeeding plantacions: the lord make it like that of New England: for wee must Consider that wee shall be as a City upon a Hill, the eies of all people are uppon us.

The belief that America serves as a moral paradigm, or model, for the rest of the world has continued to be a major component of the national consciousness. It validates the conviction, characteristically important to Americans, that the United States is morally qualified for world leadership. It accounts for the persistent tendency of Americans, both privately and in their official capacities, to take a judgmental attitude toward the policies and actions of other countries, as well as—it must be emphasized—of the United States itself. In this way, American foreign policy has always been infused with moral considerations that constitute self-imposed restraints, on the one hand, on the exercise of American power and, on the other, on U.S. capacity to deal effectively with countries and situations deemed morally wrong.

The second major element in the American sense of mission has been the conviction that America provides a unique example of a free and democratic society that can and will be copied by the rest of mankind, which for countless ages has lived under tyranny and oppression. Gaining prominence during the Revolution, this view was proclaimed by Thomas Paine in *Common Sense* with the ringing words:

> We have it in our power to begin the world again. A situation, similar
> to the present, hath not happened since the days of Noah until now.
> The birthday of a new world is at hand.

Jefferson prophesied that the United States was "destined to be the primitive and precious model of what is to change the condition of man over the globe," and Lincoln believed it to be "the last, best hope of earth."[6] Indeed, from the late 18th century until today, the conviction that the United States is a sociopolitical paradigm for other nations has been an ineradicable characteristic of the American self-image. It accounts for the widely prevalent implicit assumption that the natural process of world evolution is toward the kind of freedom and democracy exemplified by the United States, to which all other nations are believed to aspire and which they will sooner or later achieve. Woodrow Wilson, the leading 20th-century proponent of this expectation, proclaimed in an Independence Day speech in 1914, that:

> America has lifted high the light which will shine unto all generations
> and guide the feet of mankind to the goal of justice and liberty and
> peace.

Wilson's belief that the United States and its allies were fighting in World War I "to make the world safe for democracy" and his crusade for a postwar settlement that would hasten and protect the inevitable attainment of national independence and democratic institutions by all peoples of the world are too well-known to require further description here. Deeply imbued with the moral imperatives of the Protestant ethic, Wilson was convinced that America's greatness lay in actively fulfilling its vocational mission to lead the way to the worldwide system of free and democratic states that alone could ensure universal and enduring peace.

The Wilsonian conception of America's world leadership responsibility also includes the third—and latest to be developed—component of the sense of national mission. Increasingly after the Civil War, the economic growth and technological advances of the United States fostered the conviction that, in addition to being a moral and sociopolitical exemplar, America serves as a model of technoeconomic progress that other nations can and must emulate if they are to improve the health and living standards of their people. This progressist element in the American sense of mission is closely related to the sociopolitical component through the belief that U.S. economic achievements are inseparable from the advantages of political democracy and the private enterprise system. The faith in the heuristic power of American technoeconomic progress helps to account for the strong rationalistic and prescriptive tendencies of American foreign policy, particularly in the decades since World War II.

All three components of the American sense of mission reached their fullest and most dynamic expression in the postwar period of America's preeminence as a superpower, when it had an unprecedented opportunity to provide leadership for reconstituting the international system. In such circumstances, even Adlai Stevenson, that most intellectual of postwar political leaders, could sum up America's sense of national purpose with the proclamation that:

> God has set for us an awesome mission: nothing less than the leadership of the free world. Because He asks nothing of His servants beyond their strength, He has given to us vast power and vast opportunity. And like that servant of Biblical times who received the talents, we shall be held to strict account for what we do with them.[7]

Since the late 19th century, the growth of U.S. power absolutely and relative to the changing power relationships and national aspirations of other states in the international system has fostered the shift from a passive to an active mode of realizing the American sense of mission. For, it is important to recognize that *both isolationism and interventionism are expressions of the sense of mission.* In the first, pure America keeps itself aloof from the contamination and dangers of entangling alliances with European powers and from involvement in the affairs of other regions. Such isolation enables it to fulfill its destiny as "a City upon a Hill" to which "the eies of all people" are irresistibly drawn in their eagerness to learn how to build a virtuous, free, democratic and progressive society. In the active interventionist mode, America is not only a shining example for others to copy but it also has the inescapable obligation to impel them into doing so and to protect and assist them until they, too, become self-reliant. Thus, the difference between isolationism and interventionism lies not in the nature of America's mission but in the manner of expressing it.

The shift to the active mode in the course of the 20th century also reflects the increasing positivistic conviction of mastery over nature and society. Two kinds of American positivism can be distinguished: the common-sense popular variant unique to the United States; and the typical technocratic elite-group form similar to—but much more intense and self-confident than—that of Europe. The sets of ideas and attitudes subsumed under each are not mutually exclusive, and both are held in some degree by most Americans despite the inconsistencies between them.

The set of beliefs and expectations comprised in common-sense popular positivism is rooted in the interactions between the achievement-oriented immigrants seeking a new beginning in America and the pressures of the harsh environment, natural and social, in which they had to realize it. Until the 20th century, a distant central government gave them little more direct help than the maintenance of law and order, the financing of some "internal improvements" (i.e., roads, canals and railroads), and the granting of free homesteads. Hence, both the earlier British settlers, with their Protestant ethic of worldly asceticism, and the later immigrants from Ireland and continental Europe, with their more traditional sense of organic social solidarity, had to rely upon themselves, their families and neighbors for assistance in coping with natural and man-made dangers and difficulties. The continuing success of individual, family, local-community and ethnic-group efforts fostered, and was in turn sustained by, the widespread development of such personal characteristics as self-reliance, self-confidence, optimism, versatility, flexibility and activism. In the main, the problems faced by Americans were solved by ordinary people using common sense—that is, generally shared ideas about nature and society that seemed plausible in the light of everyday observation and experience and which, when acted upon, appeared to be pragmatically effective.

By its nature, common-sense popular positivism tends to be anti-intellectual, suspicious of professionalism, and regarding theory and sophisticated technique as unnecessary.[8] Sometimes a reaction against and corrective to the arrogance of technocratic positivism, this pervasive popular anti-intellectualism constitutes one phase of the more general paradoxical American attitude toward professionalized scientific and other intellectual skills; in its opposite phase, it exalts them as the infallible means of human mastery over nature and society. Popular anti-intellectual feeling contributes a major element to the "know-nothingism" and paranoid fears of alien and domestic conspiracies that are generated in the United States during periods of acute national frustration and uncertainty.

Common-sense popular positivism leads to a sentimental parochialism in American ways of thinking about the nature and future development of the international system. Americans are inclined to believe that their common-sense conceptions and practical abilities have universal validity, transcending all environmental differences, natural and social. What works in America is bound to work elsewhere; and, if it doesn't, the fault lies not in the American common-sense approach but rather in the peculiar customs of other nations and peoples. Moreover, the latter's difficulties could be readily overcome by self-reliant, outgoing, cooperative interpersonal relationships, like those that played so important a role in America's own development. Thus, the international system as a whole and the national societies comprising it could be transformed into a peaceful, prosperous and progressing world community of nations through common sense, warm people-to-people relationships, and mutual goodwill and helpfulness.

In contrast to common sense and the natural capabilities of ordinary people, technocratic elite-group positivism relies upon professionals trained in highly technical skills, using specialized scientific knowledge, preferably working in teams, and endowed with the most advanced equipment and ample funds. It is rooted in the interactions between the strongly rationalistic element in the American cultural heritage and the unprecedented technoeconomic accomplishments of American society.

American rationalism became significant in the course of the 18th century, and its earliest formulations depended heavily upon the philosophical and political ideas of John Locke. In the United States as in England, and partly stimulated by British developments, the empirical aspect of Lockean rationalism evolved into *laissez-faire* utilitarianism by the early 19th century. Reflecting and in turn reinforcing the growing technological and managerial orientation of U.S. economic and social development, American rationalism and individualism were also enriched by the influences first of French positivism and later of German social science. Educated Americans found increasingly congenial the confident faith of Condorcet, Comte, Saint-Simon and later French positivists in the redemptive power of science and technology and in rational methods of organization and operation. They were similarly drawn to the professionalism, systematic methodology, and emphasis on social structure and process common to the different German historical and socioeconomic schools.

These diverse influences helped to form that turn-of-the-century generation of Americans who, in the period 1890–1920, created or modernized the professional schools and physical and social science departments of American universities, and thereby shaped the educational determinants of the subsequent development of technocratic positivism in the United States. Most of this key generation were raised on the farms or in the small towns of rural America, where their ways of thinking were rooted in common-sense popular positivism, with its activist self-reliance, democratic equality, and Protestant sense of vocation and moral responsibility. Their formal education imbued them with the individualism, pragmatic empiricism, and commitment to private initiative characteristic of the prevailing Anglo-American liberalism. Finally, many of them engaged in postgraduate study at European—especially German—universities, where they were attracted to the professionalized systematic conception and organization of research in the physical and social sciences. Their efforts transformed American higher education into the institutional means for developing the elitist professionalism needed for scientific and technological advancement and for inculcating into successive generations of students the sense of personal responsibility to apply the new knowledge to raising productivity, improving social welfare, and mitigating inequalities and injustices.

The development of these attitudes and ideas both reflected and helped to foster other institutional changes—above all, those in the economic system. It was in the United States that the rationalization of economic organizations, operating methods and decision-making techniques first became prevalent in the burgeoning industrialization of the late 19th and early 20th centuries. This rationalization and accompanying professionalization made the new corporate forms of industrial enterprise the main institutional source of demand for the graduates of the transformed American universities. And, in turn, the ways of thinking inculcated by the occupational experiences of managing and working in rationalized industrial organizations further stimulated the positivistic convictions of the professionally trained elites.

Characterized until the mid-20th century by private enterprise and minimal government, the United States experienced a much slower growth, rationalization and professionalization of its governmental institutions as compared with its economic system. Unlike that in continental Europe, the evolving technocratic positivism in the United States was not identified with the all-powerful paternalistic state nor was the central government expected to play the predominant role either in developing the necessary scientific knowledge or in applying it to transform society. It was not until the great depression of the 1930s discredited the notion that private enterprise would automatically assure "the greatest good for the greatest number" and World War II generated unprecedented needs for technological research that the U.S. government became the major instrument for social progress and the main supporter of scientific advancement.

Contemporary American technocratic positivism differs from the varieties that soon also began to emerge in the continental countries not only in its earlier and more extensive development but in its more single-minded emphasis on the rational element in accounting for individual decisions and hence for the social process as a whole. The notion of atomistic "economic man"—acting as an isolated individual in a fully conscious manner in accordance with cost/benefit calculations— has been fundamental in the predominant schools of Anglo-American economics and, with a broader conception of the interests involved, in theories of political and social motivation and behavior. This has been true regardless of whether the operation of rational interest is conceived to be a voluntary, conscious weighing of costs and benefits, or whether it is believed to be mechanistically determined, as in behaviorist and physiological psychologies, economic determinism, Marxism, and so forth. Even when, at the turn of the century, Americans began to be influenced by German social science, they were inclined to select from its complex conceptions of social organization and process those aspects that could readily be combined with their own atomistic, rationalistic explanations of social behavior—resulting, for example, in the institutional school of economics and the pressure-group theory of American pluralism. It was not until the mid-20th century that American social scientists came to appreciate the deeper implications of the sociology and psychology of continental theorists, most notably of Emile Durkheim, Max Weber and Sigmund Freud. In contrast, postwar continental positivists have tended to neglect these seminal thinkers in favor either of Marxian or Jungian distortions of their ideas or of European versions of simplistic rationalism.

Nonetheless, despite the flowering of American sociology and political science in recent decades, narrowly rationalistic conceptions of decision making and behavior continue to predominate in explanations of the nature of the international system and in the formulation of U.S. foreign policy. They constitute an elite-group rationalist parochialism that parallels the popular sentimental parochialism and strengthens its effects. A major manifestation of this influence may be seen in the design of America's world-transforming mission, analyzed in the next section, that was implicit in U.S. foreign policy during the postwar period.

The effects of redemptive and positivistic modes of perception and conception are reinforced by another American cultural characteristic —the sense of guilt. An essential element in the Judaeo-Christian tradition of moral responsibility and redemptive activism, the sense of guilt reflects and stimulates the self-awareness of Western civilization and its propensity for self-criticism and self-condemnation. Strengthened by the judgmental imperatives of the Protestant ethic and the strong achievement orientation of American society, the sense of guilt inclines Americans to attribute their failures not to the unrealism of their goals but to the inadequacy of their efforts. So long as America's mission is unaccomplished, the sense of guilt helps to motivate the striving to fulfill it.

Thus, although in terms of national-interest objectives, U.S. foreign policy has been on the whole more successful since 1950 than that of the Soviet Union, this achievement gives small satisfaction to many Americans, especially among the elite groups. They do not evaluate U.S. performance by the contributions of American foreign policy to the frustration of Soviet ambitions, the security and prosperity of Western Europe, and the advances, even though modest, of some Asian, African and Latin American countries. Rather, they tend to judge the success of American foreign policy by their sense of forward movement toward the millennial goal of a secure, peaceful and progressing world order. In the absence of rapid and demonstrable progress, they are generally inclined to regard U.S. foreign policy as at fault.

American guilt feelings over failure to fulfill the sense of mission are the psychological equivalent of Russian feelings of insecurity and anxiety over the risks and uncertainties of decentralized initiative, self-responsibility and dissent. Although operating in different ways and on different aspects of personality formation and social behavior, both sets of feelings contribute to the lag in adjusting perceptions and conceptions to the changing realities of the international system. They differ also in their effect: the American sense of guilt constitutes a self-restraint on the exercise of U.S. power; in contrast, Russian anxieties tend to relax restraints on the exercise of Soviet power.

The beneficial effects of the sense of guilt are enhanced by certain other characteristics of American society and culture that continuously operate to improve the realism of its perceptions and conceptions. Chief among them are the substantial freedom for criticism and dissent, the widespread habit of public discussion and debate, and the pragmatic attitude—the willingness and ability to submit ideas to empirical tests and to distinguish between prescriptions that work and those that do not. However, American pragmatism is more effective in bringing about modification or abandonment of unrealistic domestic policies and programs than it is with respect to those aimed to control developments abroad. Within the United States, where they are subject to scrutiny not only by the participants and others whose interests are affected but also by the Congress, the communications media, the experts and public-spirited citizens, it is sooner or later impossible to conceal or ignore the fact that policies and programs are not fulfilling their purposes or the expectations regarding them. But, information about events outside a country's borders is necessarily more fragmentary, superficial and unfamiliar. The serious deficiencies in knowledge of external developments, therefore, are more readily filled with perceptions and conceptions dictated by the sense of mission and the faith in the efficacy of reason. Nonetheless, within this intrinsic limitation, the pragmatic American attitude sooner or later exerts an important reality-orienting influence on U.S. behavior in the international system.

U.S. Foreign Policy and World-Transforming Mission in the Postwar Period

Since World War II, America's sense of redemptive mission and positivistic faith in the power of reason and the inevitability of progress have constituted closely related and mutually reinforcing elements in the U.S. conception of its role in the international system. The sense of mission infuses the pursuit of national interest with powerful psychic energy and provides the moral justification for it. Positivism fosters the belief that America's world-transforming mission is practicable— indeed, sooner or later irresistible—because reason is assumed to formulate the ends to be sought and the means for doing so. Thus, the conviction that U.S. aims and policies are morally mandatory and operationally attainable generates the "inner certitude and objective certainty" that are psychologically essential for impelling action to achieve national goals.

As it took shape during and immediately after World War II, official U.S. policy envisaged "One World" comprised of large, medium and small states brought into existence by the principle of self-determination, respecting each others' sovereign independence, governed by increasingly democratic regimes dedicated to improving the welfare of their people, and conducting mutually beneficial economic and cultural relations with one another on a nondiscriminatory basis. The "collective security" and peace of such a system would be maintained by the rational interest of all, the sense of responsibility of the leading nations, and the willingness of its members to allow the United Nations—then being established—to settle disputes among them and to prevent aggression anywhere in the world. Owing to the growing democratic character and rising welfare concerns of the states composing it and to the effectiveness of the United Nations, the system would inevitably evolve toward universal peace and plenty under the rule of law. In the words of Secretary of State Cordell Hull:

> . . . there will no longer be need for spheres of influence, for alliances, for balance of power, or any other of the special arrangements through which, in the unhappy past, the nations strove to safeguard their security or to promote their interests.[9]

In the years from 1947 to 1950, however, the conviction became nearly universal in the United States that the major—indeed, to most people, the sole—obstacle to the achievement of the kind of world order projected by Secretary Hull was the unreasonable refusal of the Soviet Union to cooperate in bringing it about. Americans generally were convinced that the benefits of the U.S. design for a peaceful international system were self-evident and, therefore, every enlightened and responsible nation could not fail in its own interest to help achieve it. Moreover, with the progress of science and education and the spread of welfare-oriented democratic governments, the ranks of such rational nations would irresistibly grow. Thus, the inevitability of progress would guarantee the evolution of a rational world order.

By actively working against this American conception and by pressing for its own incompatible world-transforming goal, the Soviet Union was believed to be manifestly behaving in an irrational and perverse manner. Hence, Soviet intentions and actions were perceived as the gravest threat to the peace and freedom of all enlightened nations, collectively designated by the United States as the "Free World"—that is, the "One World" minus the communist portion which, it was hoped, was only temporarily following an aberrant course. As the largest, wealthiest, and most powerful member of the "Free World," the United States was obligated both morally and in its own interest to block the designs of the Soviet Union while seeking to achieve proximate objectives (e.g., a European political and economic union, an integrated "Atlantic Community," accelerated "development" in Asia, Africa and Latin America) that would be consistent with and steps toward its own ultimate goal. By so doing, it was argued, time would be gained during which Russian society, if not the existing Soviet dictatorship, would make sufficient economic and political progress to recognize its own rational interest in the kind of international system projected by the United States and to cooperate in working toward it. Thus, the American use of such "special arrangements" as alliances, deplored by Secretary Hull, was justified not as a relapse into the futile expedients of "the unhappy past" but as a means of hastening the advent of a secure and prosperous future.

This characterization of U.S. foreign policy in the postwar period as no less world-transforming in aim than that of the Soviet Union may seem at variance with the still prevailing American and European interpretations which, in greater or lesser degree, regard the Soviet Union as the revolutionary aggressor and the United States as the defender of the *status quo*.[10] However, the conventional view and that presented here can be reconciled, at least in part, because they relate to different aspects of American foreign policy. It was revolutionary in the sense that its goals and expectations envisaged a transformed world order, which Americans assumed to be attainable and confidently anticipated achieving in the not too distant future. In its behavior, the United States was conservative in situations where it felt that it had to contain actual or suspected Soviet and local communist initiatives. Its mode of operation was generally—although not invariably—constrained by its own democratic and humanitarian values, including its commitment to the principle of national self-determination. Nevertheless, in certain major respects, even its actions were then and have continued to be revolutionary in intent. This characteristic is most marked in the strong American pressure during most of the 1960s for unprecedented political and economic unification in Western Europe, discussed in Chapter V, and in the still persisting U.S. effort to use the influence believed to be provided by its financial and technical assistance to Asian, African and

Latin American countries as the means for radically transforming their traditional agrarian authoritarian societies into modern industrialized democratic states.[11]

The latter objective of U.S. foreign policy reflects not only the American sense of mission but also the simplistic rationalism and parochialism of American positivism, in both its technocratic and common-sense forms. Throughout the 1950s and '60s, the great majority of U.S. government officials and opinion leaders in the various elite groups were confident that the rational interest of Asian, African and Latin American countries would be sufficient, if supplemented by U.S. financial and technical assistance and sound economic advice, to ensure that they would rapidly increase their rates of economic growth and distribute their incomes more equitably. Even though the frustration of these positivistic expectations is now beginning to lead to the recognition that economic growth is an inseparable part of a much larger and more complex process of sociocultural change, U.S. officials still tend in practice to discount the effects of political, other social-institutional, cultural and psychological factors. The implicit conviction is still widespread in the United States that the inertia of traditional institutions and the conservatism of customary values and behavioral norms in Asia, Africa and Latin America can be easily overcome by technoeconomic prescriptions. Americans continue to believe that the deep conflicts of interest among particularistic groups in the countries of those regions and their competing conceptions of desirable national goals and priorities can be readily resolved by rational discussion and compromise.[12]

A similar faith in the power of reason underlies American ideas about the capacity of a worldwide system of sovereign states to preserve peace and increase justice and welfare among and within its constituent members. Conceptually, this view involves a distortion of Locke's distinction between a state of nature and a state of war, explained in Chapter IV. Although Locke himself recognized that a peaceful state of nature was essentially precarious and often deteriorated into a state of war,[13] most Americans implicitly regard peace as the only normal condition of a system of sovereign states. In consequence, they are inclined to believe that war is an irrational aberration, an exceptional and unnatural occurrence, that could be eliminated. In effect, the American view is that, if they wish, sovereign nation-states are collectively capable of acting rationally to keep the peace in the same way as, and more successfully than, individuals can in a state of nature. Hence, there is no serious, much less intractable, obstacle preventing enlightened nations from living together in enduring harmony, cooperating among themselves to advance their mutual welfare, and settling disputes by directly negotiated compromises or by the decisions of an international tribunal. Recognizing their rational interest in making such a system work, the great majority of nations would be willing to apply sanctions against the few threatening to break the peace or refusing to abide by international arbitration.

Implicit in this conception was the parochial assumption that, as the most rational of all, the United States would set the standard for proper international conduct and hence would never be subjected to coercive sanctions. For this reason, too, the most important test of the moral worth and pragmatic effectiveness of the policies of other countries, as well as of the United Nations as the organizational means for achieving a rational world order, was the extent to which they conformed to the aims and prescriptions of the United States.

The unrealistic expectations of rapid sociocultural advances in Asia, Africa and Latin America and the distorted Lockean conception of the power of rational interest to create and maintain a peaceful and progressing international system have been congenial to most Americans not only because they are expressions of the ingrained positivism of American culture. They are also, in a sense, projections onto the international system of certain distinctive characteristics and historical experiences of American pluralism.

In the highly differentiated American society, groups, organizations and individuals have an unusual degree of freedom to protect and advance their own interests. Except when conflicts become too severe, as during the Civil War, the late 1960s, and other periods of deep national division, the disintegrative effects of competition and controversy are controlled by institutional factors, such as democratic political processes and the judicial system, and by cultural characteristics, such as the universalistic values and behavioral norms fostering concern for the national interest or public good and willingness to abide by the results of political and judicial decisions. Along with the high degree of voluntary cooperation and mutual aid developed among the individuals, families and ethnic groups who settled in America, these sociocultural experiences generated the belief that supervening coercive power, latent or manifest, was not essential to the maintenance of order and peace. For these reasons, Americans have been inclined to assume that, like their own pluralistic competitive society, an international community composed of many different states pursuing conflicting interests could also resolve disputes by negotiated compromises or by voluntary submission to the determinations of an arbitrational agency without power to enforce its decisions.

In these and other ways, the sense of redemptive mission, the parochial assumption that the people of other societies and cultures naturally think and act like Americans, and the optimistic confidence in the efficacy of rationalistic panaceas tend to obscure American perceptions of external realities and lead often to the choice of objectives that are unattainable, and means of policy that are ineffective. Usually, these cultural influences help to distort or to reinforce the more familiar and conspicuous institutional elements shaping American foreign policy and external behavior.[14] However, although in the main conducive to unrealistic expectations and exaggerated reactions, their effects are also in some respects beneficial.

For example, in contrast to Soviet policy, which is tactically flexible and strategically inflexible, U.S. policy tends to be the reverse. Because its world-transforming mission is conceived in very general terms, the United States is not committed to a specific long-range objective—as the Soviet Union is to the establishment of communist regimes in other countries—but rather to a general direction of development for the international system and its constituent nations. True, missionary zeal and positivistic ways of thinking continually generate demands among American opinion leaders that the President and the State Department define more specifically the goals of U.S. foreign policy, that they adopt a strategy for assuring world peace or a plan for "developing" Asia, Africa and Latin America, that they draw up a blueprint of the orderly and progressing world community of nations which the United States is seeking to achieve. Such panaceas are commonly prescribed for reversing the setbacks of foreign policy and are believed by many opinion leaders to be infallible means for rallying other nations in support of the world-transforming aim of the United States. And, inability—as well as unwillingness—to comply with these demands makes the State Department unpopular. Nevertheless, the resulting flexibility with respect to long-range objectives has usually been a source of strength for U.S. foreign policy.

The characteristics that permit strategic flexibility also operate to produce a comparative tactical rigidity. As exemplified by the long, frustrating involvement in Indochina, it is difficult for the United States rapidly and easily to make substantial changes in its short-term aims and immediate means of policy. This inflexibility tends to sustain unrealistic expectations, whose inevitable disappointment in turn inhibits policy changes—for example, recognition of China—and discredits policies and programs—for example, support for the United Nations and for foreign aid—that, when realistically conceived, are desirable and practicable.

The frustration of unrealistic expectations from time to time produces certain ambivalent reactions. On the one hand, American pragmatism usually leads sooner or later to painful adjustments to unwelcome realities. On the other hand, the resentment engendered by the necessity to adjust is sometimes relieved, not by the more common guilt reaction of condemning the inadequacy of U.S. efforts, but by projecting the blame onto the deficiencies of others, abroad and at home. Such scapegoats include the malign irrationality of the Soviet Union, the short-sighted self-seeking of European or Asian allies, the incompetence of the State Department, the imperialist machinations of the President, the Pentagon and Wall Street. The shattering of American hopes for "One World" by the developing cold war and the invasion of South Korea led in the late 1940s and early 1950s to the McCarthyite hunt for the "traitors" responsible. Twenty years later, the frustrations of the Indochina War and the economic disputes with Western Europe and Japan brought to a head the growing disillusionment of Americans over the analogous failure to achieve certain of the most cherished objectives of U.S. policy in the 1960s—especially victory in Indochina, and the ready conformity to American wishes by wealthy trading partners assumed to result from their supposed identity of interests with the United States. The disappointment of these expectations underlay the feeling of many American opinion leaders and legislators in the late 1960s and early 1970s that the United States was "Uncle Sucker," who trustingly sacrificed his own welfare to assist ungrateful allies, and it helped to motivate their willingness to support the protectionist measures pressed by special-interest groups.

As in the case of the Soviet Union, a variety of internal and external factors were operating in the late 1960s to moderate the redemptive activism of American foreign policy. The external changes are described in the next chapter. Essentially, they have had the effect of narrowing the freedom of action of the United States and reducing the scope for seeking—much less for attaining—world-transforming goals. In contrast to the postwar period of cold war, therefore, U.S. behavior in the international system is today and for the foreseeable future more circumscribed and its redemptive activism is accordingly less conspicuous. Yet, despite these changes, external circumstances are likely from time to time to encourage or permit more activistic manifestations of its persisting expectations that, through U.S. leadership or inspiration, the international system would be transformed.

The internal developments moderating the redemptive activism of U.S. foreign policy are themselves expressions of the American sense of mission to perfect society. As embodied in the new nationalism, they comprise the growing insistence on more rapid achievement of an expanding range of domestic goals, and the focusing of elite-group and popular attention on the economic difficulties and social conflicts thereby engendered. For, the pressures to achieve domestic goals have involved not only struggles over resource allocations but also the more complex and less tractable issues of race relations and the violence and fear they generate. Moreover, these domestic distractions and controversies have been magnified by the diversion of resources from internal goals and the inflation resulting from official misconceptions of how long the war in Indochina would last. But, although these internal concerns moderate the redemptive activism of the United States in the international system, they do not remove it. As many Americans pressing for domestic reforms assert, one of the benefits of achieving their objectives will be that, by eliminating the grave deficiencies of its own society, the United States can once again resume its necessary role as a moral, sociopolitical and technoeconomic exemplar for the rest of the world.

Nor does the reaction against U.S. involvement in Indochina portend an end of American redemptive activism in the international system or even a new period of isolationism. Most Americans opposed to the Indochina intervention have not been advocating termination of active U.S. participation in world affairs or abandonment of America's world-transforming responsibility—which, in any case, has an isolationist as well as an interventionist mode. Rather, their criticism has been directed to the manner in which official policy makers have been trying to carry out this mission.[15] Much of the controversy regarding Indochina has been premised on a conception of U.S. intervention as a means to the larger objective, not as an end in itself. Many who oppose it have insisted that it is an action incompatible with achievement of America's world-transforming goal; those who support it have claimed that it is an essential measure for realizing the same end. Particularly by young people, U.S. actions in Indochina and elsewhere are indicted as contrary to the ideals of freedom and democracy and the principle of national self-determination. U.S. foreign policy is condemned for transgressing the very values that constitute America's qualification for its redemptive mission, and the means for carrying it out can be justified only by restoring their moral purity.

Not only is the persistence of the sense of mission demonstrated in these ways but, as explained in Chapter II, American society and culture are becoming more, not less, technocratic. Hence, in the foreseeable future, there are bound to be periods of greater, as well as of lesser, U.S. interest and activity in the international system. In the course of the 1970s or early 1980s, it is likely that a new administration would once more be impelled to "get America moving again" by pursuing new forms of activist policies shaped by the conviction that technocratic panaceas can readily solve the problems not only of the United States but of the rest of the world as well. Indeed, because significant analogies can usually be seen only with the benefit of hindsight, there is even a reasonable probability that the United States would again be tempted by its sense of mission to intervene in situations elsewhere in the world to which, at the time, the lessons of Korea or Indochina would appear inapplicable.

The Consequences for World Politics of American and Soviet Perceptions and Conceptions

Different though they have been and continue to be in specific content, American and Soviet views of the nature of the international system and of their respective roles and responsibilities in it have been equivalent in their significance for world politics. In the early postwar years, each projected a revolutionary scheme for a new world order that would once and for all eradicate the deficiencies of the existing system. And, even today, the United States still expects to transcend the limitations of nationalism by means of rational interest; the Soviet Union still professes to eliminate the evils of imperialism through socialism. Each persists in seeing the other as the main enemy blocking the achievement of its own world-transforming goal. Each still believes that its conception of the future will sooner or later prevail—that of the United States assured by the power of reason and the inevitability of progress; that of the Soviet Union guaranteed by the inexorable dialectic of history. Each continues to be convinced that it has a special responsibility for providing leadership, ideas and resources by virtue not only of its size, strength and vital interests but also of its moral and institutional superiority.

Most important of all for the actual course of world politics, each set of perceptions and conceptions has tended to validate the other. That is, those expressed in the intentions and behavior of one superpower are among the most important facts of life that the other superpower has to take explicitly into account in formulating its foreign policies and actions. Yet, neither superpower's perceptions and conceptions have constituted an accurate reflection of the real nature of the international system and of the factors and forces working in it. Obviously, both sets have had to conform to reality at least to the extent necessary to yield a minimum degree of operational effectiveness to actions based on them. However, even such pragmatic tests have in significant degree reflected performance expectations partly engendered by their respective sets of perceptions and conceptions.

The interaction between these two sets has constituted, in effect, *a logic of mutual distrust* in the relationships between the United States and the Soviet Union. Each superpower has tended to be impelled to a "worst-case" interpretation of the intentions and behavior of the other. Each has been convinced that it must be prepared in every way—militarily, politically, economically, scientifically—to prevent the unfavorable outcome for itself of every situation believed to be capable of affecting its interests, directly or indirectly, because it could not afford the risk of giving the other the benefit of the doubt, much less of taking the other at its word. Hence, each move of the other had to be countered; indeed, every event anywhere in the international system had to be scrutinized to ascertain how it might benefit the other or adversely affect itself. Both superpowers have been, in the vivid 17th-century prose of Thomas Hobbes (more fully quoted in Chapter IV) "in the state and posture of Gladiators; having their weapons pointing, and their eyes fixed on one another . . . and continuall Spyes upon their neighbours." It was these characteristics of the developing relationship between the United States and the Soviet Union that, in the late 1940s, were summed up in the term that came to be the conventional name for the postwar period—"the cold war."

Thus, the logic of mutual distrust operated throughout the 1950s to make the United States and the Soviet Union invest with cold-war significance virtually everything that happened in the international system. Regardless of where they were located geographically and of the substantive relevance of the problems involved, nearly all conflicts of interest and other disputes among nations, their constructive or disruptive initiatives in world and regional politics, and the political and economic crises within independent countries and colonial areas were regarded by each superpower, if not always as caused by the machinations of the other, at least as likely to result in unacceptable advantages to the other side. In consequence of this universalism, all regions of the planet were conceived to be actual or potential battlegrounds in the cold war.

The worldwide scope of the cold war was expressed in the efforts made by each superpower during the 1950s and early 1960s to protect those states initially on its side and to persuade or pressure as many other nations as possible to align themselves with it. For, both superpowers were convinced that, in the apocalyptic struggle in which they were engaged, all nations would sooner or later be compelled to join one side or the other. The notion of the unavoidability of alignment was applied especially to the new nations of Asia and Africa. Both superpowers believed Asian and African countries to be so inchoate and weak as to limit their capacity to determine their own destinies solely to the choice between the American and the Soviet roads to a peaceful and progressing world order.

Accordingly, each superpower organized a system of mutual defense arrangements, which it dominated. The Soviet Union bound the East European communist countries to it in the Warsaw Pact and concluded mutual defense agreements with China, North Korea, North Vietnam, Cuba, and several newly independent noncommunist states in Asia and Africa. The collective defense treaty system organized by the United States was even more extensive, embracing 40-odd countries. In addition to the United States, the system comprised the West European nations and Canada in the North Atlantic Treaty Organization (NATO); the Latin American countries in the Rio Treaty; Australia and New Zealand in the ANZUS Treaty; Japan, the Philippines, South Korea and Taiwan in separate bilateral pacts; and several of the foregoing plus Thailand and Pakistan in the Southeast Asia Treaty Organization (SEATO). The United States also entered into an informal relationship with the Central Treaty Organization (CENTO), consisting of Iran, Pakistan, Turkey and the United Kingdom.

The uncompromising and universal dichotomy of the cold war meant in effect that, for many countries, world politics tended to be internalized and domestic politics to be externalized. This phenomenon was not, of course, a unique characteristic of the postwar international system—it was already manifested during the interwar years when relations with Nazi Germany and the Soviet Union strongly affected the domestic politics of other nations. However, in the postwar period, the extent and intensity of this phenomenon were greatly magnified. In the case of the United States, although foreign policy issues had occasionally in the past been interjected into domestic politics, they tended in the two decades after World War II to dominate the sense of national purpose, strongly influence the allocation of economic resources, warp the perception of many purely domestic events, generate extremist and paranoid reactions on the right and the left of the political spectrum, and disrupt the bipartisan support for U.S. cold-war policy. In lesser degree, a similar internalization of world politics occurred in Western Europe. In Asian, African and even Latin American countries, too, the question of whether politicians and parties were for or against the United States, or the Soviet Union, or unaligned was a major influence on the course of domestic politics. Conversely, the internal developments and problems of many countries, especially the newly independent nations, were externalized as issues in the cold war.

The question naturally arises as to whether the protection and advancement of national interests *per se* are sufficient in themselves to explain the worldwide scope, the revolutionary character, the immense diversion of resources from pressing domestic purposes, and the intensity of feeling that typified American and Russian foreign policies and activities in the postwar decades. National security interests and prudent suspicions would probably have been sufficient to bring about many of these actions—especially the development and preservation of mutual nuclear deterrence, the NATO and Warsaw Pact alliances, and probably the system of overseas bases and naval forces. They would very likely also have led to some U.S. and Soviet interventions in the affairs of other nations to achieve or prevent significant shifts in relative power positions, including perhaps the as yet hypothetical case of preserving access to critical raw materials unobtainable elsewhere. They might even have justified large-scale subsidies to other countries to enhance U.S. or Soviet influence and, as in the Marshall Plan, to prevent deteriorations that might directly or indirectly have threatened the armed truce existing between the superpowers.

But, neither superpower has been compelled by its vital interests to go beyond such limited objectives and policies, and instead to seek to transform the very nature of the international system and of the societies of the nations comprising it. Certainly, their political security did not require them to do so; nor did their economic welfare. The list of strategically important commodities for which each superpower is wholly or substantially reliant upon imports is quite small and, for most of them, substitutes—no doubt more expensive or less satisfactory—are or would soon be available. Otherwise, the continental size and diversity of their economies make both superpowers nearly self-sufficient with respect to imports and not dependent upon exports to sustain adequate rates of economic growth. Hence, while it might have been economically

advantageous for them to dominate other regions, they were not compelled to do so by economic needs.

It is, I believe, only by taking into account not simply their national
interests and desire to preserve their paramount positions but also their
senses of world-transforming mission and the resulting logic of their
mutual distrust that it is possible to explain American and Soviet attempts during the postwar period to transform the international system,
their convictions that they knew how and had a special responsibility to
do so, and their determined and costly efforts to accelerate and guide
economic, social and political changes within other countries in all parts
of the world.[16] Moreover, although the nature of world politics is different today than in the postwar period, both sets of perceptions and
conceptions will continue to help shape American and Soviet actions in
the international system over the foreseeable future. For, the ongoing
power of their senses of national identity and purpose not only is derived from deep roots in the historical development of American and
Russian societies but is also continually renewed by the existing psychosocial processes of personality and attitude formation and the current
institutional constraints in the two countries.

The Prospects for World Peace
and War

THE PURPOSE of this chapter is to assess the probabilities of a world war—that is, a full-scale nuclear war between existing or prospective superpowers—in the foreseeable future. The reason is that such a war is the only way in which events outside the Atlantic region would be likely to change *fundamentally* the societies and cultures of the Western nations. True, other kinds of developments in Eastern Europe, Asia, Africa, and Latin America could seriously affect the political security or economic well-being of North America and Western Europe. But, however important they might be in degree, such effects would not be likely to alter in kind the political and economic systems of Atlantic nations during the remainder of the century. In contrast, a world nuclear war would certainly change very radically, if it did not completely destroy, their main institutional and cultural systems.

This relative autonomy of the Atlantic region is derived from the self-instigating, dynamic character of Western civilization, whose momentum is great enough to ensure that, at least for another generation and probably longer, it will continue to be predominantly self-determining, except for the contingency of a world nuclear war. Moreover, this probability is enhanced by the likelihood that, except perhaps for Japan, none of the other societies and cultures on the planet will in the next decade or so experience such far-reaching political, technoeconomic, scientific, philosophical or aesthetic developments as to make their influences of determinative importance to the Atlantic nations.

After briefly noting certain characteristics of any system of independent nation-states, the chapter first sketches the changes within and among the countries comprising the worldwide system that brought to an end the postwar period of cold war. Then, the distinctive features of the present period are identified, and the more and the less probable ways in which they could develop over the next two or three decades and their likely effects on the prospects for a world war are analyzed. However, there is a difficulty in discussing these subjects that is inherent in the systems-model approach followed in this book. Because such a conceptual framework interprets the course of events as shaped by complex interactions within and among national, regional and worldwide systems, the analysis in this chapter presupposes not only the material presented in the preceding chapters but also that still to be examined in the succeeding ones dealing with Western Europe and the Atlantic region, respectively. Similarly, the analysis in those chapters is premised on the projections of the worldwide system presented here. For this reason, some repetition and cross referencing are unavoidable.

The Nature of an International System

The characteristic that most markedly distinguishes a system of independent states from the social system of a nation is the fact that the former contains no sovereign authority with the power and the right to adjudicate disputes among members and enforce its decisions, whereas the latter does. Each independent political entity has norms of behavior governing the relationships among its constituent individuals and groups and means for deterring and punishing those who transgress its customs and laws. Although certain carefully defined forms of violence (dueling, boxing, hunting, for instance) are permitted in modern societies, it is a valid generalization that the legitimate use of force in them is restricted to collective authorities, who decide in what circumstances and manner to apply it. Thereby domestic order is maintained and peace is made the normal condition within national societies.

The situation is otherwise in an international system. Because it lacks a sovereign authority, war itself and the continuing threat of war are conditions as normal to it as peace. Such a situation has traditionally been characterized as a "state of nature." Probably the best known description of this concept is that of Thomas Hobbes, the 17th-century political philosopher, who wrote:

> ... without a common Power to keep them all in awe, they are in that condition which is called Warre; and such a warre, as is of every man, against every man. For Warre consisteth not in Battell onely, or the act of fighting; but in a tract of time, wherein the Will to contend by Battell is sufficiently known.
>   ... in all times, Kings, and Persons of Soveraigne authority, be-

cause of their Independency, are in continuall jealousies, and in the
state and posture of Gladiators; having their weapons pointing, and
their eyes fixed on one another; that is, their Forts, Garrisons, and
Guns upon the Frontiers of their Kingdomes; and continuall Spyes
upon their neighbours; which is a posture of War.[1]

Such a state of armed vigilance, of war without fighting, has characterized the international system in certain periods, for example, during the
cold war of the 1950s. However, John Locke maintained that conditions need not be continuously bellicose. He found that

. . . the state of Nature and the state of war, which however some men
have confounded, are as far distant as a state of peace, goodwill,
mutual assistance, and preservation; and a state of enmity, malice,
violence, and mutual destruction are one from another. . . . Want of
a common judge with authority puts all men in a state of Nature;
force without right upon a man's person makes a state of war both
where there is, and is not, a common judge.[2]

Whether in any particular period the international system will ex-
hibit Hobbes' mutual vigilance or Locke's mutual goodwill or whether
it will possess other distinguishing features depends essentially upon
the configuration of power capabilities and the objectives and actions
of the states comprising it at that time. These relationships reflect two
sets of determinative factors. The first consists of the relative geographical locations and the relative magnitudes of the populations, economic
resources, political effectiveness, and military strength of the states in
the system. The second are the ways, analyzed in preceding chapters, in
which each member state conceives its own national interests and goals
(including, most basic of all, self-preservation) and perceives the aims,
capabilities and behavior of the others. Member states seek their objectives and advance their interests unilaterally, and multilaterally
through alliances, spheres of influence, international organizations and
the less formal kinds of associations and responsibilities they undertake and choose to observe.

Both the configuration of power relationships among the states comprising an international system and their societies and cultures change
over time, and hence the characteristics of systems change. To take
account of such differences, political theorists have devised various
classifications for international systems. One, based on the aims of the
leading nations, distinguishes between "moderate" systems, which tend
to restore and maintain equilibrium among major powers, and more or
less extremist "revolutionary" systems, in which one or several great

powers try to change fundamentally the configuration of the system and even the societies of the states comprising it. Another classification reflecting the structural elements identifies "bipolar" systems, in which two preponderant powers and their respective allies compete for hegemony or confront each other in more or less unstable equilibrium, and "polycentric" systems, in which several major powers or separate groups of states, usually pursuing limited aims or marginal adjustments, interact with one another. These two classifications are compatible and, as in the preceding sentence, may be combined to provide finer distinctions.[3]

The Ending of the Postwar Period

The postwar period of the cold war can be regarded as embracing roughly the two decades from the late 1940s to the mid-1960s. Its beginning and terminal dates cannot be more precisely expressed because neither the onset of the period nor its ending were signaled by specific dramatic events, such as the outbreak of a war, nor were they immediately evident to the participants, including the two superpowers. Nonetheless, as early as the second half of the 1950s changes began to occur in the international system that in the mid-1960s were to bring the postwar period to an end. In part, they consisted of modifications in the bilateral relationship of the two superpowers. More significant, however, were the new developments elsewhere in the international system that, both directly and through their influence on the bilateral relationship, began to have increasingly important effects in the course of the 1960s.

Changes in the Bilateral Soviet-U.S. Relationship

The Cuban missile crisis of 1962 may be regarded as the dividing event marking the change in the bilateral relationship between the Soviet Union and the United States. Prior to it, there was a succession of more or less direct confrontations between the superpowers either in Europe (as in Berlin) or in other parts of the world, which involved the possible resort to military force, including even nuclear weapons. Since then, neither side has provoked similar confrontations, not even over U.S. involvement in Indochina or the Soviet occupation of Czechoslovakia and intervention in Egypt. In such potentially critical situations, both the United States and the Soviet Union have tried to avoid behaving in ways that would involve them in direct confrontations requiring the threat or the use of military power.

This moderating of the cold war reflected in part the nuclear parity that emerged in the course of the 1960s. Throughout the postwar period, the United States possessed both first- and second-strike superiority over the Soviet Union, although the latter's retaliatory capabilities were sufficient to deter the former from resorting to nuclear attack. While American officials expressed suspicion in the late 1950s that nuclear parity had been reached, the Soviet Union did not in fact attain such equality with the United States until a decade later. Nevertheless, in the circumstances, neither wished to put to the test the other's determination to use its nuclear weapons in a critical situation of direct confrontation. Each superpower continued after 1962 to promote its design for a new world order but by means that would not involve a direct threat to use force against the other—a situation characterized by Khrushchev as "competitive coexistence."

The change in the relationship between the two superpowers both helped to bring about and was itself fostered by parallel changes in the relationships of each superpower with its respective allies and with the nations of Asia, Africa and Latin America. In addition, as explained in the preceding chapter, the internal problems of the two superpowers have played significant roles in these interacting changes.

Developments in the Soviet Alliance System

The difficulties in the relationships between the Soviet Union and its allies basically reflect the fact that the countries of Eastern Europe have in the past century and a half been more influenced socially and culturally by Western Europe than has the Soviet Union. Hence, they experienced a much more substantial movement toward a liberal order than did the latter. Their attitude toward the West has not been conflicted, like that of the Russians, but has always been predominantly admiring and emulative. In consequence, liberalizing tendencies once again manifested themselves within these countries during the 1960s. In addition, the ethnic nationalism and hatred of colonial rule that impelled their independence movements in the 19th and early 20th centuries have continued to play a major role in their external relationships. In the course of the 1960s, these factors led them to seek, if not always to carry out, reforms—more thoroughgoing and advanced than those of the Russians—both to meet the pressures of their people and to reduce the disadvantages of their detailed central planning and of their one-sided economic relationships with the Soviet Union, imposed on them by the Russians along with their communist regimes.

Although perforce they have been obedient members of the Warsaw

Pact, the East European countries have been restive under Soviet direction of their foreign policies and reluctant to divert substantial amounts of their resources, needed at home, to assist other communist states, pro-Soviet or anti-American regimes, and unaligned nations in Asia, Africa and Latin America. Their desire and ability to express these divergent interests reflect and reinforce their economic recovery from the effects of World War II and the increasing popular acceptance and resulting greater self-confidence of their communist regimes.

In consequence of these and other changes in the course of the 1960s, Soviet influence over most of the East European countries has become less assured. The Soviet Union has not been able to count on the automaticity of their support for its foreign policies and on the alacrity of their acceptance of its advice on their domestic policies that existed in the postwar period, when they were in the fullest sense Soviet "satellites." Indeed, the measure of the uncertainties of the Soviet Union's control over its East European allies is the fact that it had to use force to effectuate its will in East Germany, Hungary and Poland during the 1950s and in Czechoslovakia as late as 1968, thereby also again intimidating the others. The strains and divergences within the Warsaw Pact system not only weaken the unanimism of the Soviet hegemony but also exacerbate the Soviet regime's anxieties that its domestic control might be undermined.

The cohesiveness of the Soviet alliance system has been further diminished by the emergence of China as a messianic rival for the leadership of communism's world-redemptive mission. This competition has been expressed in substantial differences in tactical doctrine regarding the ways of achieving the mission. In addition to rivalry for communist leadership, it also reflects long-standing conflicts over the longest common frontier on the planet and the prospect of dominating Eastern and Southern Asia. In the course of the 1960s, the split with China compounded the constraints and frustrations of Soviet foreign policy, especially its relationships with other communist nations and with communist parties in noncommunist countries.

The Loosening of the Atlantic Alliance System

A far-reaching decline in the cohesiveness of the NATO system and in U.S. influence over its West European allies also became manifest in the course of the 1960s. The first critical postwar decade of economic reconstruction and political recovery in Western Europe was succeeded by a second decade of unprecedented economic growth and rising living standards. The nature and implications of the ensuing changes in attitudes and capabilities in Western Europe will be discussed in detail in Chapters V and VI. Here, their influence in shaping the characteristics of the current period of world politics are briefly noted.

Contemporary West European nationalism is the product of the economic recovery and unprecedented prosperity of these countries during the 1960s and, more fundamentally, of the emerging technocratic transformation of their societies sketched in Chapter II. One major consequence has been the powerful and pervasive feeling at all levels of the population that European resources should be devoted to increasing domestic welfare, in the broadest sense of the term, and the corresponding focus of attention on the expanding institutions of the nation-state.

The new nationalism is also expressed in the more complex attitudes of the younger political and economic elites—as distinct from the literary and philosophical intellectuals—analyzed in Chapter V. The younger politicians, businessmen and technocrats share the popular preference for raising living standards and improving the conditions of life, which they are confident of being able to do through the advancement of science and the rational management of the economic system. Their technocratic positivism, validated by their contributions to the economic growth and prosperity of Western Europe, in turn reinforces their self-confidence, activism, sense of national purpose, and impatience of external direction and restraint. These developing elite-group attitudes underlay the increasing European resistance during the 1960s to U.S. preponderance in the Atlantic system, the refusal to adopt American proposals for integrated Atlantic and European defense arrangements, and the rejection of the U.S. conception of Western responsibilities in other parts of the world.

The United States, too, has become more nationalistic in recent years. The growing pressure on resources, the deepening domestic divisions, and the frustration of American expectations regarding the behavior of other nations generated a more direct and self-interested interpretation of U.S. objectives abroad by the late 1960s. This shift also reflected the changes since the 1950s in the relative military position of the United States *vis-à-vis* the Soviet Union and in its relative economic position *vis-à-vis* Western Europe and Japan. As to the first, the Soviet attainment of nuclear parity by the late 1960s has already been noted. As to the second, Chapter VI will explain the gradual emergence in the course of the decade of the European Community as the world's largest trading entity and the extraordinary growth and international importance of the Japanese economy. Although the U.S. gross national product still exceeds that of Western Europe as a whole, the United States no longer has either the decisive technological lead or the determinative economic influence that it enjoyed in the Atlantic region and the international economy during the postwar period. In consequence of these developments, the United States has been increasingly unwilling to use its resources and to forgo existing or additional benefits for the sake of other countries or of the world polity and economy.

These changes in Western Europe and the United States were reflected in the gradual divergence of interests and attitudes regarding the purposes and operations of NATO that became evident in the course of the 1960s. Although differences existed between the smaller and larger members and among several of the major European countries, the most marked and significant divergence was between the United States and the other allies as the direct Soviet menace to Western Europe noticeably diminished. In turn, the divergence deepened Western Europe's ambivalence and frustration over its participation in the U.S. alliance system.

All West European states, including France, continued to rely upon the U.S. nuclear deterrent as their ultimate protection against the residual Soviet threat to their independence, and they were thereby relieved of the only compelling necessity to develop their own nuclear capabilities. But, they also began to feel both less certain about the reliability of the U.S. guarantee and more eager to have an influential voice in determining NATO policy. The attainment by the Soviet Union of an intercontinental nuclear missile capacity, then the U.S. development of second-strike capabilities, later the Soviet achievement of nuclear parity and the related unilateral shift of NATO strategy by the United States from "massive retaliation" to "graduated response" successively weakened—although they did not destroy—Western Europe's faith in American willingness to use nuclear weapons in its defense rather than only in retaliation for a Soviet attack on U.S. territory. This uncertainty underlay both the French interest—exaggerated but not created by de Gaulle's ambition—in developing its own national nuclear force, and the resistance of France and initially of other potential nuclear powers in Western Europe to the nonproliferation treaty promoted by the superpowers. At the same time, however, except for France and, in diminishing degree, the United Kingdom, all were unwilling to devote substantial resources to developing their own nuclear capabilities and even to achieving the conventional force levels agreed upon in NATO. Yet, such measures were essential prerequisites for obtaining the influence on NATO policy making that the West Europeans increasingly desired.

These frustrations, combined with the effects of the receding Soviet threat and Western Europe's growing economic integration and prosperity, led in the course of the 1960s to a loosening of the NATO alliance and a substantial decline in U.S. influence over West European policies. But, unlike the Soviets, whose security would be threatened by East European defections, Americans were neither impelled by their anxieties nor permitted by their moral standards to enforce their will upon their European allies. Instead, the United States tended on the whole to be forbearing, even under occasional severe Gaullist provocation, in part for tactical reasons. In part, too, U.S. restraint was fostered by the conviction that the growing European refusal to follow American leadership was only temporary and would sooner or later be overcome by the inevitable recognition of Europe's rational interest in supporting U.S. conceptions of new European, Atlantic and global systems of order.

Changes in Asia, Africa and Latin America

The 1960s also witnessed the virtual completion of the great postwar decolonization movement that brought into existence 60-odd new nations in Asia, Africa and the Caribbean area. The newly independent countries, as well as the few older nations in Asia and Africa, have tended to conceive of themselves as active rather than passive members of the international system, and—with some exceptions—as neutral with respect to the superpowers. In consequence of their active participation during the postwar period, world politics became truly global for the first time in human history.

Certainly the most conspicuous form of contemporary nationalism is that of the new and old nations in Asia and Africa. Extensive studies of Asian and African nationalism have led to differing views regarding its nature, duration and significance for the international system. My own interpretation has been presented elsewhere in detail and need only be summarized here.[4]

In essence, Asian and African nationalism is a central feature of the transitional process through which the societies of these regions are now passing. It is the major form of their search for new cultural and national identities to replace their traditional senses of identity shattered by the impact on them of Western society over the past 150 years. It expresses itself during this phase of their development primarily in their external relationships with one another and in their conflicting desires to emulate and reject the West, especially the United States. In part through these external encounters and experiences, Asian and African countries are developing a sense of their own unique identities. In time, these strengthening senses of national identity and purpose will integrate more effectively the many incongruent and competing groups, institutions and values comprised in these particularistic societies and transitional cultures.

The Asian and African search for national identity is so central an aspect of the development process that it impels these countries to reject control by any outside power, especially one that is disproportionately strong or rich, even though they may be dependent on its military support and economic assistance. Thus, even the pro-Soviet or pro-Chinese ruling elites in several of these countries are more nationalist than communist and hence have refrained from instituting communist regimes. Those that have—North Korea and North Vietnam, as well as Cuba—resist domination by Moscow or Peking, however much they may concur in Soviet or Chinese designs for transforming the international system. Moreover, the credibility of communism as a panacea for easing and accelerating the development process has significantly declined since the 1950s in consequence of the persistent economic difficulties of the Soviet Union and other communist states and the continuing Russian reliance upon repressive totalitarian institutions.

An equally significant, if less outwardly manifested, nationalism characterizes the countries of Latin America. Except in the Caribbean, the region consists of old nations that achieved their independence more than a century and a half ago. Although they have long since developed senses of cultural and national identity, the Latin American countries still feel the need to safeguard vigilantly their national independence and sovereignty *vis-à-vis* the United States, and they continue to regard one another with considerable jealousy and competitiveness. Their fear of the United States underlies their rising criticism of U.S. private investment and their search for alternative economic and political relationships with Western Europe, Japan and the Soviet Union.

As noted in Chapter III, the prevalent view during the 1950s in the United States, the Soviet Union and Europe was that few, if any, Asian and African countries could avoid alignment, voluntarily or perforce, with one or the other superpower. Although this expectation proved unfounded, the United States and the Soviet Union, as well as the former colonial rulers in Western Europe, have persisted in exercising greater or lesser degrees of influence in Asia and Africa. Moreover, the external initiatives of the countries of these regions, the inconsistencies within their societies, and the precariousness of their political regimes continue to be important sources of disorder and crisis in the international system. In accordance with the logic of their mutual distrust, the superpowers regard such situations as either opportunities to advance or threats to their own interests.

The Transitional Phase of Limited and Blurred Bipolar Competition

In the course of the 1960s, the effects of the disorders and of the constructive and disruptive initiatives of Asian, African and Latin American countries have combined with the changes in the relations between each of the superpowers and its respective allies to confuse and diffuse the clear-cut global dichotomy of the cold war that had dominated the 1950s. In turn, this development reinforced the effects of nuclear parity and mutual deterrence in moderating the bilateral relationship between the United States and the Soviet Union and in fostering the policy of competitive coexistence.

Not only did the two superpowers cease to be the only important determiners of world politics by the late 1960s, but their fiats were no longer self-enforcing within their respective alliance systems, as in different degree they had been during the 1950s. Each could spur on an ally or client state to more intense or extreme action along the lines of its own desired behavior, as the Soviet Union did in the preliminaries to the Egypt-Israel war of June, 1967, and to the India-Pakistan conflict of 1971. But, the converse was no longer generally the case. Once an ally or client state decided to do or not to do something of great importance to itself, usually the application, or credible threat, of force by a superpower has been needed to drive it to the contrary action.

The various types of nationalism that became manifest in the course of the 1960s made the international system much more complex than it was in the early postwar years. The bipolarity resulting from the global competition of the superpowers continued to be a major element in the structural configuration of the system. But, other sets of re-lationships expressing different national orientations increasingly constituted additional structural elements that introduced new and redistributed existing stresses and strains among nations and regions. Or, to shift to a dynamic metaphor, it is as though the currents in the sea of world politics, previously determined by the rival centripetal attractions of two gigantic whirlpools, were now being partly broken up and diverted by numerous smaller eddies and crosscurrents no less dangerous to navigation.

The net effect of these trends and countertrends of the 1960s' was that the activities and accomplishments of the United States and the Soviet Union became more circumscribed and incomplete than they were in the postwar period. The United States and the Soviet Union were more and more constrained to pursue limited aims and to do so increasingly by indirect and restrained tactics rather than by direct and uncompromising means. Indeed, in the course of the 1960s, they tacitly recognized each other's spheres of influence. Combined with their common fear of provoking a nuclear war, this situation meant that the superpowers were as concerned with maintaining the *status quo* as with changing it. In these circumstances, it is not surprising that the respective radical critics of the two superpowers accused them of playing conservative roles in world politics. The Declaration of Principles agreed upon by the Soviet Union and the United States at the 1972 Moscow summit conference represented formal recognition by the two superpowers of the changes imposed by the nature of the new period on their mutual relationship and on their behavior in the international system.

Thus, as the characteristics of the new nationalism became increasingly evident, the international system ceased to be revolutionary. For much of the postwar period, there had been both a need and a widespread desire to reconstitute a world order different from that of the prewar years. In contrast, by the late 1960s, there was and will probably continue to be a viable system that functions reasonably well. Regardless of the many ways in which people believe it should be improved, the system has nevertheless become acceptable in at least minimum degree to the ruling elites in most countries. Other than by positivists and Marxists who expect eventually to transform it, the existing international system is rejected today only by those political leaders and intellectuals in Asia, Africa and Latin America who seek a redistribution of wealth on a planetary scale. However, few—if any—of these countries are likely to be willing or able to merge their sovereignties in a united effort to force such a redistribution; and, even if they could, it is very doubtful that their combined strength would be adequate for the task. Hence, for most countries, as well as perforce for the dissidents, acceptance of the existing international system is implicit in the different forms of nationalism that motivate its constituent members.

For these reasons, the international system combines basic stability with considerable insecurity. It is stable in the sense that the threats to its constitution—to the fundamental configuration of power relationships—are held in check by the constraints imposed by the superpowers on their own behavior and on that of the few other nations desirous and capable of overthrowing it. The system is insecure in the sense that it is prone to recurrent crises and disorders generated by the external initiatives and internal problems of any one or several of its constituent states. This situation is analogous to that of Latin American countries in the 19th and early 20th centuries, when their social institutions and class relationships were stable but their political regimes were precarious, liable to frequent palace revolutions and *coups d'état* that neither resulted from nor caused significant changes in their political systems. The insecurity of the international order compounds the intrinsic difficulty of predicting future *events;* its stability permits the kind of projection of its future *trends* made in the next section.

The Emergence of a Balance-of-Power System in the Period of the New Nationalism

The new period of world politics that began to emerge in the early 1960s is still too young for any one of its characteristics to be perceived as sufficiently outstanding to be used as its designation. Nevertheless, the resurgence of nationalism in the course of the 1960s has had a most important effect not only on internal trends within the Atlantic countries but also on their external behavior in both the regional and the worldwide systems. Nationalism, too, is a major characteristic of the external relations of most nations in other parts of the world. Hence, until a better name for it can be discerned—perhaps only with the benefit of hindsight—the current period of world politics may be designated as that of the new nationalism. The qualifying "new" is important because the kinds of nationalism prevalent today differ from the aggressive, xenophobic type that predominated in the 19th and early 20th centuries. They also differ from one another, as explained in the preceding section.

Owing to the changes leading to and, in turn, fostered by the new nationalism, the development of the international system during the 1970s and '80s is likely to be shaped increasingly by two trends. The first is the greater differentiation of the hierarchy of states as some nations grow to certain crucial levels of population, industrial capacity, political effectiveness and military strength. The second is their con-comitant tendency to play more and more independent, active and directive roles in the international system.

The Concept of Proto-Superpower

The nations already approaching, or in the foreseeable future capable of achieving, this status may be termed—for want of a better word—*proto-superpowers*. For inclusion in the category of proto-superpowers, countries have to meet certain minimum qualifications. They must already possess, or be likely to have by the end of the century, populations of 100 million or more and political regimes able to make and carry out the domestic and foreign policies required for an active and influential role in world affairs. They need to have industrial systems and technoscientific capabilities at least equal to those of China today. Highly industrialized nations, such as in Western Europe, can be classified as proto-superpowers even though their populations may be smaller—say, around 70 million—because of the greater power implications of their more advanced technologies, larger and more diversified industrial capacities, and experienced and effective governments. By this flexible definition, the imminent and potential proto-superpowers would be France, Germany (even with-out unification with East Germany) and the United Kingdom in Europe; China, Japan, India and Indonesia in Asia; and Brazil and Mexico in Latin America.

The significance of these proto-superpower qualifications is that they provide a country with the ability to develop a serious nuclear capability. Unlike the two superpowers—for whom the maintenance of paramountcy and the logic of mutual distrust work toward maximi-zation of offensive and defensive armaments—proto-superpowers need to have only a fairly secure second-strike capability. To obtain such a nuclear sanction and means of blackmail, it would not be necessary for them to have industrial systems as large and diversified as those of the United States and the Soviet Union. In order to develop the required nuclear weapons and intercontinental means of delivery, certain minimum levels of industrialization and of technoscientific skills are, of course, essential. But, as the case of China indicates, they are attainable by quite a number of countries considerably less in-dustrialized than those of North America and Europe.

The first question is whether the nuclear capabilities of the proto-superpowers would be taken seriously by the two superpowers, and the major consideration is the latter's perception of the former's ability to inflict an unacceptable amount of destruction. This means, in effect, that the proto-superpowers would have to have a credible ability to penetrate the nuclear defenses of the superpowers to the minimum necessary degree—a test of their electronic, computer and missile technologies rather than of their nuclear skills *per se.* Granted the continued development of offensive and defensive systems by the superpowers and the vast economic and technoscientific resources at their disposal, substantial efforts would be required of the proto-superpowers to achieve and, thereafter, maintain the minimum necessary size and sophistication of their means of delivery. However, such efforts would not be beyond their economic resources and technoscientific abilities, especially in view of the likelihood that they would have access in various ways to some at least of the results of the much larger and more advanced research and development programs of the two superpowers.

The crucial question is not whether the proto-superpowers could actually destroy the superpowers in a nuclear attack. If relative nuclear offensive and defensive abilities were the sole considerations involved, as they are in game theory and other formalistic analyses of military strategy, the disparity would be decisive and the status of proto-superpower, as defined in this projection, would be theoretically impossible. In real-life situations, so many nonmilitary factors affect decision making, and often operate in such contradictory ways, that the minimum credible retaliatory capability of the proto-superpowers could provide leverage on, and means for blackmailing, the superpowers *as long as it constitutes a threat and is not used in actual combat.* Political, psychological and moral considerations would constrain the superpowers to accord much more significance in diplomacy to the nuclear capabilities of the proto-superpowers than they would merit in warfare. At a minimum, the superpowers would be more chary of putting pressure on the proto-superpowers than they would on other nations. As the examples of China and Gaullist France foreshadow, the more active and independent a role the nuclear proto-superpower plays, the more seriously would it be regarded by the superpowers; conversely, as the example of the United Kingdom in the 1960s shows, the more docile an ally it is, the less would it be able to influence the policies and actions of the superpowers.

In this projection, therefore, the question of whether the United States and the Soviet Union would have to take seriously the nuclear threats or blackmail of the proto-superpowers relates not to the ability but to the *willingness* of these countries to allocate resources for developing a reasonably secure second-strike capability and to the purposes for and ways in which they would use their resulting power in the international system. And, because their economies are smaller than those of the United States and the Soviet Union, their motivation to divert proportionally greater resources to this purpose would have to be strong.

In turn, the motives for allocating the necessary resources will be determined not only by the opportunities to advance their interests or the threats to their security perceived in the international system but also by the pressures generated within their societies and the dramatic designs of their cultures. Hence, one or more of them would have to be impelled by a sense of mission to begin actively to pursue world- or region-transforming goals. Or, at the least, some or all of them, cooperatively or competitively, would have to insist upon playing independent roles alongside the two superpowers in the management of the international system. In either of these ways, therefore, the future evolution of world politics—and especially the manner in which the present period might end—would in this projection depend upon the revolutionary or independent behavior of the new proto-superpowers and the reaction of the old superpowers.

China as a Revolutionary Proto-Superpower

Chinese civilization has been in the past, and is capable of again becoming, one of the most creative on the planet. That it has rivaled Western civilization in this respect is demonstrated not only by its own achievements but also by the current capacity for adaptation and growth of its offshoots in other parts of Asia, as well as of Japan and Korea, whose past development was deeply influenced by China. Since the communist revolution, the traditional conception of its superior status of "Middle Kingdom" has been reinvigorated and made dynamic by imported Marxist messianism. Hence, despite its time of troubles in the 1960s and its current economic and military limitations, China has to be included in the category of prospective proto-superpowers owing to its sheer size, its demonstrated technoscientific capabilities, and the developmental potentialities of its great historical culture.

Indeed, the influence of Marxist messianism and the effectiveness of its communist regime have already imbued China with a sense of redemptive mission which, although hitherto largely inward directed, could become much more outward oriented in the foreseeable future. The ruling communist elite conceives of China as the world-revolutionary rival of the Soviet Union, and the Chinese provide assistance to nationalist and communist movements of various kinds not only in Indochina but elsewhere in Asia, Africa and Latin America. The key questions regarding China's future role in the international system are whether the nation will remain sufficiently unified and, if so, how much more expansive and aggressive its sense of mission will make it become.

With the receding of the "cultural revolution," the likelihood of a breakup of China into autonomous states seems more remote although, in view of its past history, such an event cannot be considered outside the limits of the possible. It is more difficult to estimate the probability of China's pursuing actively expansionist policies in the foreseeable future. During the 1970s, its economic limitations and the pressures to raise the living standards of its immense population would very likely continue to restrict its major external initiatives to the neighboring portions of Asia. But, if and as progress is made in mitigating domestic deficiencies, China could become an increasingly active force in world politics and the scope of its interests and ambitions in other regions could broaden accordingly. Moreover, even if China's concern remains predominantly limited to Asia, its own national security and welfare interests, the directions of expansion dictated by topography and population-density disparities, and its historical grievances and boundary disputes will lead to recurrent conflicts with the Soviet Union, India and other neighbors—as well as perhaps with the United States over the contiguous island nations—at times involving local hostilities that could threaten to escalate into nuclear war. In any event, therefore, China's behavior is likely to constitute one of the main influences in the international system in the decades ahead. It will probably confine itself to the limitations of competitive coexistence. However, there is also a reasonable probability that China's sense of world-transforming mission would sooner or later impel it to play the role of a revolutionary proto-superpower.

The European Community as an Independent Proto-Superpower

The West European nations are more likely to pursue world-transforming goals by implication rather than by deliberate intent. In effect, they would seek to play an active, independent and important part along with the superpowers in managing the international system. Moreover, the probability is greater that they would try to do so collectively through their membership in the European Community than on an individual national basis. Such a joint effort would be more powerful, and hence more influential in world affairs, and its costs would be less difficult to bear relative to the resources available.

Indeed, the development of the European Community into a full economic and political union would create an entity so much larger and more powerful than France, Germany or the United Kingdom alone that it would become another superpower rather than a proto-superpower. But, as explained in the next chapter, the likelihood that the Community will achieve a federal union in the foreseeable future is substantially less than that it will continue to maintain and improve its integration in ways which do not involve transfer of the crucial economic and political powers to supranational authorities. If so, an eventually confederal Community would develop in the course of the 1970s a sufficient degree of military coordination and foreign-policy cooperation to enable its members to act more or less as a unit in world affairs.

As explained in Chapter VI, the European Community has already attained this status in the international economic system. Not only is it the world's largest trading entity but it has or will soon have preferential association agreements with all the nonmember West European nations, virtually all black African states, and most other countries bordering on the Mediterranean. Thus, the Community now constitutes the center of a powerful world trading and investing bloc both cooperating and competing with the United States and Japan. The conception of the future organization of the international economic system implicit in European bloc formation differs sufficiently from that hitherto envisaged by the United States for the issues involved to become increasingly important sources of contention. Combined with the pressures generated by the other persisting problems in transatlantic economic and political relationships analyzed in Chapter VI, a European bloc is likely to constrain both the United States and Japan, however unconsciously or unwillingly, to form economic blocs of their own. And, the competition and conflicts of interest that would then inevitably arise among these world trading blocs would, in turn, press the Community into greater military and foreign-policy coordination in order to increase its influence in world economic and political affairs.

Hence, regardless of whether Western Europe develops into a federal or a confederal entity and whether British, French or German elites tend to predominate in it, the European Community would play an increasingly active, independent and important role in the international system. The difference between federal union and confederal integration is, nevertheless, significant in two related respects. First, as explained in Chapter V, the strength and extent in the Community of the will to become a superpower are likely to be the decisive factors determining whether it will develop in the federal or the confederal form. Second, the activism and sense of world-transforming mission of a European federal union would *ipso facto* be substantially greater than those of a looser, more heterogeneous European confederation, in which national identities and differences would remain. Nonetheless, if the Community fails to become a superpower through federal union, it would at least become the premier proto-superpower through confederal integration. For, what it would lack in size, unity and sense of redemptive mission, as compared to China, it would more than offset by its much greater economic power and more advanced technoscientific skills.

Although neither France nor the United Kingdom has yet developed a sufficiently credible second-strike capability, the military basis for the Community's proto-superpower status already exists in the French and British nuclear forces. Hitherto, both nations, and especially France, have been opposed to integrating their forces, much less to Europeanizing them. During the 1960s, the United Kingdom tried to preserve, and Gaullist France attempted to attain, the equivalent of proto-superpower status on an individual national basis. But, both efforts ended in failure. Hence, it is likely that each of these countries will sooner or later conclude that its only practicable chance of playing an independent and important role in world affairs is as the dominant influence in a confederal European Community.

Thus, it is significant that the British political and opinion leaders mainly responsible for the reversal of the United Kingdom's policy toward European unification were younger members of the elite groups. Like their seniors, they were indifferent, if not hostile, to British participation in a European political and economic union during the 1950s. However, in the early 1960s, it became apparent to many of them that neither of the alternatives hitherto pursued for maintaining an influential British world political role were effective. First, the United Kingdom by itself was unlikely soon to be able to generate enough resources to reverse the trend of withdrawal from world responsibilities while continuing also to meet popular welfare expectations. Second, such developments of the early 1960s as the abrupt Skybolt cancellation, the Nassau Agreement, the pronounced respect and preference of the Kennedy Administration for the European Community, coming in the wake of the American refusal to support the United Kingdom during the Suez crisis, meant that the "special relationship" with the United States was no longer a means by which the British could exert a significant influence in world politics through their capacity to affect the policies and actions of one of the superpowers. In these circumstances, the only other possibility was to join a united Europe, which would possess the resources required for proto-superpower status and which the experience, skill and self-confidence of British political leaders and civil servants would enable them to direct.

For France, the choice of forgoing individual proto-superpower status is more difficult owing to the Gaullist heritage. Under General de Gaulle, France went furthest of all the NATO countries in repudiating American leadership without sacrificing the ultimate protection of the U.S. nuclear guarantee, and in asserting a foreign policy often at odds with that of the United States. Indeed, de Gaulle wanted France to be treated by the United States and the Soviet Union as though it were already a major nuclear power. And, to a significant extent, he succeeded. Yet, in the face of popular insistence on obtaining broader welfare benefits than would have been consistent with de Gaulle's worldwide ambitions, even his charisma was insufficient for maintaining him in power, and the likelihood is small that, in the foreseeable future, another charismatic leader could succeed where he failed. Moreover, Germany's greater and still growing economic power and its more active role in East-West politics have been generating increasing concern in France, a development that helped to remove the latter's objections to British membership in the European Community. Thus, in time, a majority of the French elites are likely to recognize that the most realistic way in which France could play an independent and important role in world politics would be by cooperating with the British in making the European Community into a proto-superpower. And, the self-confidence of French civil servants and technocrats would foster their belief that they would be able to exercise the leading voice in a confederal Community by holding the balance between the British and the Germans.

Germany and Japan as Separate Proto-Superpowers

So far in the 20th century, Germany has made two revolutionary assaults and Japan one on the constitution of the international system. Since their defeats in World War II, both have enjoyed exceptional economic growth and prosperity; and they have been liberalizing the persisting elements of their traditional patrimonial orders to a degree unprecedented in their previous development. Moreover, throughout the postwar period, they maintained close political and economic ties with the United States, upon which they relied for nuclear deterrence of Soviet or Chinese aggression. Both were intent upon internal growth and correspondingly reluctant to devote substantial resources to augmenting their military capabilities and playing active, independent roles in world politics. For the future, Japan, whose gross national product has already surpassed Germany's, is likely to continue to grow at rates that, well before the end of the century, would make its economy larger than that of the Soviet Union and closer to that of the United States. Germany will probably widen its already significant lead over the other European nations.

Owing to their rapidly growing economic resources, the increasingly technocratic character of their societies, and the dynamism inherent in their cultures, neither country is likely to continue indefinitely under American tutelage and protection. The coming to power in both nations during the 1970s of a new generation that feels neither the guilt of Nazism or militarism nor the humiliation of defeat means new national self-confidence and assertiveness. Hence, despite persisting inward-oriented concerns on the part of the great mass of their people, the younger German and Japanese technocratic elites will sooner or later press for their countries to play more active, independent and powerful roles in world politics.

It is important, therefore, to recognize that the social relationships and attitudes in which their former senses of imperial mission were rooted have not yet passed away. Indeed, the traditional respect for and conscientious obedience to hierarchical authority of their people played no small part in the German and Japanese economic "miracles" of the 1950s and '60s. True, in both countries, postwar institutional and cultural developments have been inhibiting and beginning to displace these patrimonial characteristics, and it is likely that this process will continue. Nonetheless, basic relational and attitudinal changes commonly require a generation or two to manifest themselves fully and, until they do, there is no assurance that former self-conceptions and imperial motivations would not revive under certain domestic and international conditions.

In Germany, the trend toward more activist and positivist conceptions of internal and external objectives began to emerge with the change of administration at the end of the 1960s. Although the people generally have continued to be preoccupied with raising living standards and improving domestic institutions, the younger, more technocratic elites have been becoming increasingly concerned with

Germany's external objectives and foreign policies. Hence, the crucial question is: in which direction will they predominantly look as their sense of mission becomes more outwardly directed?

During the postwar period and the transitional phase of the 1960s, the attention and concern of the German elites were focused primarily on developing new relationships with Western Europe. And, the most probable course of events is that the westward orientation would continue to predominate in view of the extent of Germany's integration into the European Community, its fear of the Soviet Union, and the strength of its ties with the United States. In that case, Germany would be likely sooner or later to be granted the right to participate in a coordinated Anglo-French nuclear capability, which German techno-economic resources could assist importantly in developing into a credible second-strike force. The German elites would then compete with those of France and the United Kingdom for leadership of the confederal European Community.

However, the longer the European Community fails to make decisive progress toward at least a confederal arrangement, the greater the likelihood that Germany would seek to achieve proto-superpower status on a separate national basis. In that event, the possibility of a predominantly eastward direction of interest and activity cannot be excluded. Certainly, both a major historical and still strongly compelling direction for expanding German influence is toward the east. In part a reaction to gradually increasing German interest and activity in Eastern Europe, the Russian occupation of Czechoslovakia made clear that the Soviet Union would be suspicious of, and would try to limit and control, the growth of economic, cultural and political relationships between Germany and the Warsaw Pact countries. For their part, the latter—despite their memories and fears—have already become interested in expanding trade with Germany and, as they grow more restive of Soviet control, they may try to manipulate German support as a means of counterbalancing Soviet power.

Even more important is the understandable West German desire for much closer ties, if not necessarily for full union, with East Germany. Again, the younger technocratic elites have been much more flexible and imaginative in their approach to the problem of East Germany than were their predecessors under the "all-or-nothing" doctrine of the cold-war period. It is possible that, as some of the younger elites believe, a new sense of German identity and purpose might eventually evolve, which would represent a revival and further development of the traditional confederal conception of German unity displaced by

Prussian centralism in the 19th century. Effective as such a loose, heterogeneous arrangement might be in fostering closer ties between the two Germanies, there would undoubtedly be considerable difficulty in reconciling it with the growing centralization and more pervasive activity required by the technocratic state to deal with the problems of national goals and the competition for resources among them.

If Germany's direction of concern becomes predominantly eastward, these complex interests, opportunities and pressures would be likely to generate initiatives which the Germans would try—as they have already begun to do—to realize by implicit, if not explicit, understandings with the Soviet Union. The nonaggression treaty of 1970 was certainly a start in this direction. The question is how far a German-Soviet agreement could go. A mutually satisfactory, stable settlement of the issues between the two countries is difficult to conceive because the basic attitudes and interests on both sides are probably irreconcilable. Any permanent arrangement—centralized or confederal—permitting a reasonable realization of West German expectations of relations with East Germany would be bound to involve substantial diminution of Soviet control. Russian anxieties, reinforced by memories of past German behavior, and the growing restlessness of its East European allies would probably incline the Soviet Union to distrust assurances by Germany that it would not try to upset the Soviet hegemony. Hence, it is likely that the Soviet Union would continue to resist any significant expansion of German influence in Eastern Europe, as well as confederal arrangements with East Germany, however much it might wish to increase trade with Germany and to encourage German neutrality, if not help, in its rivalry with China. In such circumstances, the activist eastward-oriented German elites might well come to believe that the Soviets would be unable to deny to German nuclear power what they refused to concede to German military weakness. Depending on circumstances, Germany could develop nuclear weapons with the protective support of the United States, or clandestinely to the point where the Soviet Union would be deterred from intervening to stop it.

Japan's increasing technocratic positivism and redemptive activism are likely to become more and more outwardly directed under the stimulus of developments affecting its relations with the superpowers and proto-superpowers. Already in the early 1970s, its postwar ties with the United States have been greatly weakened by the unilateral manner as well as the substance of the U.S. *rapprochement* with China, the restrictions imposed on certain Japanese exports to the United States, and the emergency monetary and trade measures temporarily adopted by the Nixon Administration in 1971. The spur provided by these developments to a more independent and active Japanese role in world affairs will probably be reinforced by its continuing economic needs. As an island nation with a slender natural resource base, Japan is highly dependent upon imported raw materials. To assure access to sources of supply, Japan has been negotiating long-term purchase contracts with and investing in Asian, African and Latin American countries—arrangements that could serve as the basis for building a world trading bloc of its own as this process accelerates during the 1970s. The resulting competition and conflicts of interest with the other trading blocs would, in turn, stimulate Japan's desire to increase its political power in order to be taken more seriously by its rivals and its actual and potential client states. This reaction would enhance the probability of Japan developing its own nuclear capability, an outcome that would also be fostered by declining Japanese confidence in the reliability of the U.S. nuclear guarantee.

This process of becoming a proto-superpower would be decisively accelerated by the emergence of China as a revolutionary proto-superpower. Even if Japan's reviving sense of mission does not impel it to try to reestablish an East Asian hegemony, it would be likely to feel that its security and interests were threatened in proportion to the strength and expansionism of a Chinese proto-superpower dominating the contiguous mainland. Thus, in its own defense and to ensure the continued growth of its economic relations and political influence in Eastern and Southern Asia, Japan would be convinced that, despite traumatic memories of nuclear destruction, it would have to develop its own nuclear capability if it had not already done so.

Potential Proto-Superpowers and Other Nations

Elsewhere in the world, the achievement of proto-superpower status by India or Indonesia—in addition to China and Japan—could transform Asian regional politics, reducing the freedom of action and importance of the United States and, to a lesser extent, the Soviet Union, and resulting in the emergence of a shifting and precarious regional balance-of-power system. In Latin America, U.S. difficulties would be compounded by the rise of Brazil or Mexico to proto-superpower status and the acquisition of token nuclear weapons by more industrialized middle powers, such as Argentina. Brazil and Mexico have already begun to challenge U.S. leadership in minor ways, and their efforts to do so are likely to increase as their populations, economic resources and political effectiveness grow, and especially if they eventually develop the minimum credible nuclear retaliatory capability necessary for proto-superpower status. The outcome could be a bipolar regional system in which the United States and the remaining smaller client states still dependent upon its political support and economic aid would be counterbalanced by a reasonably effective coalition of larger nations.

In addition to the prospective and potential proto-superpowers, numerous other nations in the next several decades will become sufficiently industrialized to produce for themselves or, despite the nonproliferation treaty, will be able to obtain from allies and friends small arsenals of nuclear weapons and means for delivering them, especially within their own regions. Such nations might be called "middle powers." They would be distinguished from the proto-superpowers by their substantially smaller populations and economic systems and by the predominantly regional rather than worldwide focus of their external political activity. In this projection, therefore, the nuclear nonproliferation treaty and other efforts by the superpowers would be more likely to slow down rather than to prevent the spread of nuclear capabilities among middle powers.

The most obvious reasons impelling middle powers to develop or obtain nuclear weapons, as well as to continue to increase their conventional offensive and defensive armaments, would be their specific grievances and deeply rooted feelings of hostility and resentment *vis-à-vis* one another. The leading examples today are, of course, Egypt and Israel, and India and Pakistan, but similar situations involving other pairs of Asian and African nations could emerge in the future. In Asia, fear of China, and perhaps eventually of Japan, as well as of one another, and lack of confidence in U.S. willingness to defend them with its own forces, could in time make other nations feel that they had to obtain nuclear weapons of their own. Despite the "nuclear-free zone" in Latin America, the nationalistic rivalry that has engendered the arms race already evident in that region, reinforced by the desire for greater independence from the United States, could lead to nuclear proliferation if one of the potential proto-superpowers, Brazil or Mexico, or of the more industrialized middle powers, such as Argentina, develops or acquires nuclear weapons. Even in Africa in the long run, the expansionist ambitions of some nations, and their neighbors' fears, could induce them to obtain, if they could not themselves produce, at least a token nuclear armament. Finally, this process of proliferation could persuade other Western countries not members of the European Community—for example, Australia, Spain, Sweden, Switzerland, perhaps even Canada, as well as South Africa and Yugoslavia—that their own safety necessitated acquisition of a sufficient nuclear capability to deter possible middle-power aggressors.

The Balance-of-Power System

If and as the prospective and potential proto-superpowers realize more and more of their capabilities, they will do so not only to protect or advance their national interests but also because their upcoming age groups of elites want them to play active, independent and directive roles in world politics. Thus, they will increasingly insist upon sharing in the efforts of the two superpowers to manage the international system. Its character would thereby be transformed into a new worldwide version of the classical European balance-of-power model.

Such a change occurs gradually, as did that from the revolutionary cold-war period to the limited and blurred bipolar system of the transitional phase, and would similarly become fully apparent only in hindsight. The beginnings of this development were already discernible in the early 1970s in the dramatic shifts in relations among the United States, the Soviet Union and China. During those years, too, the Soviet Union, the European Community, Japan and the United States were in process of becoming the center of world economic blocs, as explained in Chapter VI. The emergence of this polycentric balance-of-power system is not more clearly apparent in part because the evolution of its military, political and economic aspects is proceeding at different rates. Thus, the Soviet Union has attained strategic nuclear parity with the United States while China and the European Community (in the British and French nuclear forces) have only regional nuclear capabilities, and Japan has none. The Soviet Union and the European Community have gone furthest in organizing world economic blocs, and Japan is likely to engage more actively in bloc formation in consequence of recent changes in its political and economic relations with the United States.

These differential rates of development of the various aspects of this polycentric system, as well as specific political and economic issues and ambitions, generate serious stresses and strains within and among the emerging blocs that both inhibit and foster the process. Such ambivalent effects may be seen in the slowness of economic and political unification in the European Community, analyzed in the next chapter, and in Japan's "agonizing reappraisal" of its role and responsibilities in the world and regional systems. Nevertheless, for reasons discussed in the foregoing pages, the probability is greater that a fully developed balance-of-power system will sooner or later become predominant than that it will not.

As in past periods of international balance of power, relationships among the great powers would, therefore, consist primarily of continuous maneuvering to bring about or prevent marginal adjustments in a system whose basic configuration would be fairly stable. Thus, international relationships will become more and more complex and fluid, characterized by shifting politicomilitary alignments among blocs, by competition among them for economic—and sooner or later for political—association with smaller countries important as markets or sources of supply of increasingly scarce raw materials, and by conflicts within and between the blocs arising from the incompatibilities of changing societies and cultures.

In consequence, most Asian, African and Latin American nations would increasingly experience three conflicting sets of pressures and constraints in their external relationships. First, their own search for national and cultural identity would continue to impel them to press for the maximum possible degree of national freedom of action. Second, their desire for nonalignment with the superpowers and proto-superpowers, as well as the positive benefits of regional economic integration and political cooperation, would maintain their interest in regional organizations of various kinds. Third, their need for trade and aid would draw them into preferential arrangements, explicit or implicit, with one or another of the world trading blocs, a relationship that would both assist their economic growth and subject them to the hegemonic influence of its leading superpower or proto-superpower—the European Community, the United States, Japan, the Soviet Union, China, and perhaps eventually Brazil, India, Indonesia or Mexico. Pulled in different directions by divergent interests and influences, most middle-size and small Asian, African and Latin American nations would suffer the resentments engendered by unsatisfactory choices and frustrated aspirations, which in turn would tend to exacerbate their internal problems. On the one hand, their regional and world-bloc affiliations would foster the stability of the international system; on the other, their external disappointments and the domestic sociopolitical difficulties accompanying economic growth would continue to make the system insecure.

The Prospects for World War

The purpose of projecting the probable course of development of the international system is to provide a basis for evaluating the prospects for a world nuclear war in the foreseeable future. Such an assessment may now be made in several ways, empirical and theoretical.

One approach is to compare the probability of a world nuclear war in a balance-of-power system with that in a bipolar system of the cold-war type.[5] The historical evidence is not conclusive because the course of world politics in previous periods—as it would today and in the future—always depended upon other major factors besides the basic configuration of power relationships in the international system. A retrospective assessment involving only the purely formal element would favor a moderate balance-of-power system. Historically, periods of bipolar competition have been characterized by more numerous and severe wars, fought for system-transforming rather than limited aims, than have moderate balance-of-power periods. The difference may be seen in the nearly continuous, exhausting hegemonic struggles first of France and Spain, next of Spain with Protestant England and Holland, and finally of France and its allies versus the Hapsburg-British coalition in the 16th and 17th centuries compared with the briefer and much less bloody wars of marginal adjustment in the 18th-century balance-of-power system. However, the relevance of the historical evidence is limited by the important differences between past and present periods, including the incomparably more rapid and immense destructive capabilities of nuclear weapons and the global character of world politics.

Considering only the form of the power relationships, the probability of a world nuclear war may be less in a bipolar system, in which only two nations possess worldwide nuclear capabilities, than in a balance-of-power system, in which many do. But, this qualification rules out a host of other variables that also have important influences on the likelihood of such a conflict. Once the other factors are reintroduced and despite the more widespread existence of nuclear capabilities, a *moderate* balance-of-power system—that is, one in which the leading nations are content to maneuver for marginal adjustments—would have a lower probability of global nuclear war than a bipolar system, in which the superpowers' sense of world-transforming mission is counterbalanced only by their own fears of mutual destruction. Such a moderate system would undoubtedly be plagued by recurrent crises as the leading nations jockeyed to balance one another, and as the initiatives and problems of other countries tempted them to fish in troubled waters and probably led from time to time to local wars. Nevertheless, the essentially nonrevolutionary character of great-power aims in a moderate system would reinforce their horror of the frightfulness of nuclear war in helping to prevent these situations from escalating into a global conflict.

In addition to the foregoing historical and theoretical methods of assessment, the prospects for a world nuclear war can also be evaluated by projecting the more and the less probable ways in which such a conflict could occur in the emerging balance-of-power system of the current period. Although no longer determinative of world politics, as during the cold-war period, the direct relationship between the United States and the Soviet Union will continue to exercise a profound and pervasive influence on the international system. Hence, the initial question is: in what circumstances would the two superpowers be likely to engage in a nuclear war?

The first possibility is that one or the other would do so by design— i.e., that it would make a revolutionary assault on the international system in a deliberate effort to achieve its world-transforming goal. As explained in Chapters II and III, the kinds of sociocultural changes needed to alter fundamentally a nation's dramatic design—its conception of its nature and destiny—occur slowly, usually over generations. In the case of the Soviet Union, its sense of mission is rooted in the inertia of traditional patterns of authority and subordination; in the case of the United States, it is sustained by the momentum of a dynamic, achievement-oriented, technocratic society. Moreover, changes in the motivating sense of national purpose are even slower to manifest themselves if external factors contribute to keeping messianic self-images alive. Activist and universalist conceptions of American and Soviet roles and responsibilities in the international system are maintained by their mutual distrust and the interest of each in preserving its superpower status. Although small, therefore, the probability cannot be excluded that either the Soviet Union or the United States would deliberately provoke a world nuclear war in the decades to come. (The possible conditions in which the United States might do so are discussed in Chapter VII.)

It is more likely that nuclear war between the two superpowers would occur not by deliberate intent but by the force of circumstances. It could be triggered by an actual or feared drastic deterioration in the position of one superpower relative to the other, by a miscalculation by one superpower of the other's responses to its initiatives, by misunderstood or inadequate communications, or by the pathological actions or impaired rationality of decision makers. One sequence of events that might lead to a war between the superpowers would be internal developments in France, Italy, Spain or, less probably, Germany involving the outbreak of civil war between pro- and anti-communist forces or the imminent coming to power of a communist regime—developments that are by no means outside the limits of the possible. In such contingencies, the United States would probably intervene, impelled by its long-standing conviction that its own security would be threatened by communist domination of any of the larger West European nations, as well as by the logic of mutual distrust. The Soviet Union would then be faced with the choice of leaving the West European communists to their own resources, perhaps with such help as it could provide indirectly or clandestinely, or of moving its forces westward to support them. In analogous situations in the past, the Soviet Union has always followed the Leninist doctrine of not risking the safety of the socialist motherland for the sake of communist movements in noncommunist countries. This would give the first alternative a greater probability in the future than the second. However, conditions within each superpower and elsewhere in the international system might at the time be of such nature as to tip the balance in favor of the latter.

Events, too, in Asia, Africa and Latin America could lead to a Soviet-U.S. confrontation that might unintentionally escalate into a nuclear war between the superpowers. The obvious source of such a sequence would be the internal problems and external initiatives of the countries of those regions. The danger would be magnified by the nuclear proliferation projected in the preceding section. Indeed, over the longer term, it is more likely that nuclear weapons would be used first in hostilities between other countries than that a nuclear war would occur initially between the United States and the Soviet Union. However, it is by no means inevitable that nuclear hostilities between middle powers, or even proto-superpowers, would escalate into a worldwide conflict involving the superpowers. Their mutual interest in preventing such a catastrophe would help to make the United States and the Soviet Union—supported by the pressures that many other countries, the United Nations, and other international institutions could bring to bear —employ varying mixtures of persuasion and threatened coercion to prevent continuation of the conflict, at least by nuclear means. Thus, even though nuclear proliferation increases the likelihood of nuclear warfare, regional and worldwide, the more probable local conflicts need not lead to global war, for they might be quickly dampened by the efforts of the superpowers and others, and even by emotional and rational reactions on the part of the participants themselves against the frightfulness of nuclear destruction.

The constraints, internal and external, that have already led to avoidance of direct confrontations and moderated the means of rivalry between the United States and the Soviet Union would certainly continue to inhibit the escalation of conflicts in Asia, Africa and Latin

America into nuclear war between the superpowers. Policy makers in both the United States and the Soviet Union are well aware of the ways by which a global nuclear war could inadvertently be precipitated and are sufficiently alert to the need for precautionary measures of various kinds—such as preserving the credibility of their mutual nuclear deterrents, improving their understanding of each other's capabilities and intentions—to lessen the chances that they would occur. And, as the 1972 arms limitation agreements show, they are prepared to restrict the deployment—and hence the cost—of new offensive and defensive weapons systems even though military research and development activities continue.[6]

The second question regarding the probable ways in which a nuclear war could occur is whether one or more of the prospective or potential proto-superpowers would make a revolutionary assault on the international system rather than participate in the marginal maneuvering of a moderate balance of power. The most likely candidate for such a role would appear to be China. Germany and Japan are other possible, although less probable, candidates. Still other possibilities in Asia and Latin America are much less probable and, in any case, they are very long term. A federal Europe, especially if dominated by Germany, or a separate German proto-superpower, might conceivably become embroiled with the Soviet Union over Eastern Europe, but a confederal Europe would undoubtedly be insufficiently united to mount such a challenge. The chance that Japan might try to build another empire by force is not great; assuming the requisite motivation, however, Japan might be tempted to do so on the mainland either by the breakup of China or in alliance with China, in which cases its antagonist would be the Soviet Union, or by trying to conquer the Pacific allies of the United States, which would involve it with the latter.

The preliminary planning and preparatory action by one or more proto-superpowers motivated by world- or region-transforming designs would mark the beginning of a new revolutionary period in the international system. Such a development would greatly increase the chance of a world nuclear war and, in any event, would result in changes of kind—and not simply of degree—in the bilateral relationship between the two superpowers and in relations between them and other states. When and how the United States and the Soviet Union would each react to the altered situation of the other would depend upon the assessment by each of the effects on its own security and interests, and of the risks and consequences of nuclear war for its own survival and postwar prospects.

In the last analysis, therefore, the answer to the second question reduces itself to whether the self-conceptions and senses of mission likely to motivate the prospective and potential proto-superpowers would be conducive to a moderate balance-of-power system. To be proto-superpowers and, hence, to be taken seriously by the two superpowers, the countries concerned must possess minimum retaliatory nuclear capabilities. Their willingness to divert the required resources from other high-priority national goals depends, in turn, on the strength of their sense of world- or region-transforming mission and on the opportunities and dangers they perceive in the international system. And, the more compelling their sense of mission, the greater the probability of a revolutionary assault on the system, all other things being equal. But, the *ceteris paribus* case does not have the highest probability. The various factors discussed in the preceding pages that would inhibit a world nuclear conflict between the two superpowers would equally affect a would-be revolutionary proto-superpower. In consequence, the likelihood that one or more proto-superpowers would deliberately provoke a world nuclear war is less than that they would not.

Nevertheless, even if none of the superpowers or proto-superpowers makes a revolutionary assault on the international system, it will continue to be highly insecure, and its balance-of-power configuration would not be all that much less susceptible to world nuclear war than the limited, blurred bipolar system of the late 1960s and early 1970s. The chances of such a conflict in the foreseeable future should, therefore, be about equal to—perhaps somewhat greater than—they were during the transitional phase of those years.

The Prospect for a Peaceful Progressing World Community

If not in a world nuclear war, how else might the period of the new nationalism end; in what other ways could the balance-of-power international system be tranformed in the decades to come? Unfortunately, it is not possible to make a projection into a future so remote because its determinative characteristics cannot yet be sufficiently discerned. Chapter VII presents an extrapolation of one possible course of development, but, it is too speculative to be included here.

Instead, it is possible to conclude this chapter by exploring briefly the question of whether the international system is likely to make significant progress in the foreseeable future toward the kind of peaceful, cooperating and prospering world community of nations that has been the abiding expectation of most Americans in the 20th century. Since the days of Woodrow Wilson, they have been taught to consider a balance-of-power system as immoral and ineffective—witness the quotation from Cordell Hull in Chapter III. Since the onset of the cold war in the late 1940s, they have been taught to think about the fate of a bipolar system in "all or nothing" terms: either the nations will learn to live in a world community under the rule of law or they will perish in a worldwide nuclear holocaust—either Millennium or Armageddon. Is such a world community a reasonably probable alternative to the possibilities projected in the preceding pages?

Among American officials and opinion leaders concerned with foreign policy and international affairs, the predominant forecast of how a world community would come about is that fundamental changes within the Soviet Union would sooner or later make the existing communist regime, or a successor, willing to renounce its world-redemptive mission and to cooperate with the United States in preserving peace and fostering international welfare. Similar changes are expected to occur in other nations—China, for instance—motivated by world-transforming designs. However, this American conception of how a peaceful and progressing world community might evolve would itself require a fundamental transformation in the nature of the international system. The revolutionary character of the change is obscured by the American propensity to assume, as explained in Chapter III, that rational interest will inevitably induce all nations, including eventually the Soviet Union and China, to support the U.S. scheme for world order.

First, although it is customary to refer to the "world community" or the "society of nations," the collectivity so designated is not a community or society in the same sense as these terms are commonly used to identify institutionally and culturally distinct social entities. Composed as they are of human beings, the nations comprising the "world community" naturally have certain common needs and interests that are met by the same or analogous kinds of institutions and values. These similarities help to make possible a greater or lesser degree of voluntary cooperation among them in dealing with international problems, and of respect for international law and custom. But, however basic or highly valued such universal human needs and interests are (such as survival in the face of common natural or social dangers, and the benefits of trade), they have never been nor are they likely soon to be as numerous and continuously pressing as are the needs and interests—and the resulting institutions and values and collective difficulties and satis-

factions—that bind together the individuals and groups comprising national, tribal, village, and other ethnically distinct societies and organic
communities. While there are individuals who think of themselves as
"citizens of the world" or as owing primary allegiance to the "brotherhood of man," the overwhelming majority of human beings derive their
sense of cultural identity from their national (or prenational) societies,
and focus their major loyalties on them.

Second, the effectiveness of institutions for adjudicating disputes depends not only on the existence of values and norms fostering voluntary
compliance with their decisions but also upon the knowledge that they,
or related constituted authorities, possess and are willing and able to
use the power required to enforce their determinations. Since the beginning of political theory in Plato and Aristotle, the crucial importance
of power has been recognized in the maintenance of peace and order
even though, as in modern democratic nation-states, it is generally
latent and does not have to be continuously manifested in the ordinary
conduct of most public and private affairs. Unlike the nation-state, the
international system contains no sovereign authority that directly or
indirectly has and is known to be willing and able to use the power
necessary to enforce compliance with its decisions. Nor, as Americans
tend to believe, can the rational interest of members of an international
system in preserving its peace and order substitute for the lack of such
institutionalized power and of reinforcing values and norms in inducing
compliance with international law and the decisions of an international
tribunal.

It is sometimes argued that, because a global nuclear war could
destroy civilization—if not all life—on the planet, the need to prevent
it constitutes a common human interest of such fundamental and overwhelming importance that it is capable of fulfilling for the international
community the integrating function that common institutions and cultures do for national societies. The analogy is not valid. On purely
rational grounds, such a common need would lack sufficient integrating
scope and power. Small nations could enjoy the benefits of an international peacekeeping institution without making the sacrifice of sovereignty and freedom of action required to join it;[7] and the large
nations, whose membership would be essential for its effectiveness,
would feel that their own nuclear forces, over which they would not
have to share control, would be better means for assuring their own
security—and certainly for achieving their ambitions. In terms of the
prospective realities of the international system, the possibility of nuclear war does not constitute so clear, continuous and concretely

focused a danger as to create the widespread conviction that survival depends upon transcending all other national interests by world unification.

Similar objections can be raised to another way in which a world community is supposed to be achieved that has become increasingly popular among Americans concerned about ecological problems. It is that the physical threat to the planet's capacity to support life will be the integrating force for creating a continuing peaceful and cooperating world order. However, the ecological threat, too, is unlikely to constitute an imminent and universal danger of critical magnitude. Serious as particular ecological problems have already become, they do not yet require an overriding mobilization of planetary attention and resources, and they will probably be sufficiently ameliorated by less drastic means long before they do. If this is so, and despite their important political implications and conditions, ecological problems will tend by their nature to be too technical to become the major theme of world politics, much less the integrating force for world unification.

Nevertheless, it is true that the most likely—perhaps the only— way in which world union might come about would be through some focused threat of such overarching and universal danger that an international system—that is, a system of separate, sovereign states—would be demonstrably unable to cope with it through its normal modes of cooperation. The role of the external Soviet menace in the movement for European union during the 1950s and early '60s will be explored in Chapter V. For the international system as a whole, however, this critical factor is likely to be supplied, if at all, by the arrival in our solar system of a technologically far more advanced form of intelligent life.[8]

The fact that the common good cannot always and unequivocally be preferred to a particular interest, especially to so basic a responsibility as national survival, makes it improbable that the United Nations, or another international institution, would be invested with the requisite police power to enforce its arbitrational decisions. In the decades ahead, the role of the United Nations is not likely to be significantly different from that which it has played during recent years. The peacekeeping and crisis-management capabilities of the United Nations cannot grow much beyond their existing forms in a system characterized by the various kinds of nationalism described above, the ambitions and fears implicit in a more differentiated international hierarchy, and the world-redemptive self-conceptions of old and new great powers.[9]

In sum, the international system, existing and prospective, is neither a Hobbesian "warre of every one against every one," in which "the life of man [is] solitary, poore, nasty, brutish, and short," nor is it a Wilsonian world community of nations, in which rational interest and democratic values preserve peace and guarantee progress. The international system is in a "state of nature," but as Locke defined it in the quotation at the beginning of this chapter. Although not subject to a superior tribunal and police force, the nations are impelled by their interests to try to keep the peace and to act with the minimum degree of mutual forbearance and cooperation necessary to maintain a reasonably effective world polity and economy. It is a community in the limited sense that nations do create common institutions to carry out important purposes that cannot be adequately served by acting separately, and they do acknowledge the legitimacy, if they do not always obey the dictates, of universal values and international law. The Hobbesians among us believe that egoistic drives and narrow self-interests are more powerful than reason and the sense of the common good. The Wilsonians among us believe that reason and the sense of the common good are more powerful than drives and interests. And, the common fallacy of both is their failure to understand that the positive and negative constraints of institutions and cultures are at least as powerful influences on human behavior as either egoistic drives or conscious reason—and must be if such a thing as human society is to exist.

V

The Progress and Prospects of European Unification

ARELY, IF EVER, in its long evolutionary history has the nation-state been under more conscious and determined attack than it was in Western Europe during the two decades following the end of World War II. Precisely in the period of its triumph, when this form of *macro* social organization had finally spread to all parts of the planet, its very survival in the region of its origin was seriously in doubt. For, not simply political philosophers and publicists but many influential politicians, opinion leaders in the main elite groups, and people generally in the European countries were convinced that the nation-state had outlived its usefulness. They believed that it would soon have to be superseded by a larger-than-national political entity if the freedom, welfare and progress of the homeland of Western civilization were to be ensured.

These convictions led in the course of the postwar period to the taking of major steps toward a united Europe that were without precedent in the modern history of the region. With the establishment in the late 1950s of the European Economic Community, explicitly intended by its six founding countries to evolve into a true political and economic union, there was a widespread expectation on both sides of the Atlantic that the passing of the nation-state was only a matter of time, at least in continental Western Europe. Yet, by the end of the 1960s, it was the future of the European union movement rather than of the nation-state that seemed the more doubtful.

Before analyzing these changes and their probable future development, three terms used extensively in this chapter and the next need to be defined. They are *integration, unification* and *union*. Although their use as broadly synonymous would be grammatically justified, this practice has resulted in considerable confusion and the dissemination of unduly optimistic or pessimistic expectations by political leaders, journalists and even some scholars. Distinctions among these terms are desirable because they point to real differences in the nature and relative probabilities of the possible economic and political relationships within Western Europe and the Atlantic region as a whole.

The term *integration* is used here in an economic sense to denote the removal of barriers to trade and capital movements among a group of countries so that, at the end of the process, goods and money move freely across national political boundaries. Integration does not involve transfers of sovereignty to supranational agencies even though the national governments concerned do lose a substantial degree of freedom of action in consequence of their mutual contractual obligations to eliminate and not thereafter restore such barriers, and of their voluntary efforts to coordinate their national economic policies. The term *unification* is used to denote a process—economic, political or military—that does require deliberate delegation of important sovereign powers to supranational authorities in one or more of these fields. The related term *union* is used as the ultimate goal of a unification movement—that is, a full federal union of formerly independent countries. Thus, in modern industrialized nation-states, political and economic unification must of necessity involve economic integration, but the converse is not true. The characteristic that distinguishes between the two processes is the degree of supranationality, that is, the extent to which the sovereignty of the individual nation-states participating in them is delegated to, or otherwise acquired by, superordinate authorities.

European Union and Atlantic Partnership in the Postwar Period[1]

European union was far from being an invention of the post-World War II period. Proposals for the uniting of Europe were made time and again throughout the centuries of the rise of the nation-state. Nevertheless, it was only in the decades after World War II that this recurrent dream of poets, philosophers and kings began to be translated into reality—and not by the short-lived military conquests of a Napoleon or a Hitler but by the more enduring voluntary actions of sovereign national governments, actively or passively supported by their people.

The Postwar Inadequacy of European Nation-States

A theory of unification has to account not only for the circumstances in which a group of nations agree to unite but also for those in which they are unwilling to do so, and it needs in addition to provide a basis for predicting how they will behave in the future. An hypothesis that meets these requirements is that the determining factor is the sense of the adequacy of the nation-state for fulfilling the functions believed essential for their basic survival and welfare by the politically active portions of the population, especially the elite groups.[2]

For the first time in their modern history, a sense of the inadequacy of the nation-state was widespread in the continental West European countries in the latter part of World War II and was intensified by the difficulties of the immediate postwar years. Several elements contributed to this growing conviction.

The first was a strong retrospective sense in the continental countries of the failure of the European nation-state system during the first half of the 20th century. The senseless slaughter of World War I and the subsequent ineffectualness of the political and economic arrangements established by the Versailles settlement; the interwar rise and triumph of Italian fascism and German Nazism; the great depression of the 1930s and the inability of national economic policies to prevent or overcome it; the ease of Nazi conquest at the outbreak of World War II, the shame of German occupation, and the humiliation implicit in having to be liberated by the Anglo-Americans and the Russians; finally, in Germany, the guilt for Nazi atrocities and the trauma of total defeat—these and related experiences of the period 1914–1945 undermined traditional continental confidence in the superiority of European culture and the effectiveness of European institutions.

Superimposed upon this sense of past failures was the manifest inability of the nation-state to cope with the problems of the immediate postwar years. It would be difficult to say in which dimension—economic, political or military—continental Europeans felt that the inadequacies of the existing national systems were greater or more dangerous for their security and welfare.

Owing to the destruction and disruption of the war, a series of relief programs financed by the United States was started even before the German surrender. These emergency efforts were replaced in 1948 by the European Recovery Program (ERP)—the Marshall Plan—which aimed within a four-year period to rebuild European productive capacity to the prewar level and, through capital investment and technical improvement, to lay the foundations for continuing increases in productivity and output. Yet, although the ERP achieved its goal, reaching prewar production levels even before its scheduled end, the general view on both sides of the Atlantic was that lagging productivity, inadequate competitive ability, restricted economic opportunities in small rigidified national markets, nondynamic entrepreneurial attitudes, liquidation of overseas investments, worsening terms of trade, and other economic changes adverse to Western Europe would persist for the indefinite future, resulting in technological stagnation, inflation, balance-of-payments deficits, and continued need for American aid.

The difficulty of coping with existing economic problems and the pessimism regarding the future were intensified by political instability and uncertainty on the continent, especially in France and Italy. Barely able to suppress street rioting and other outbreaks of violence, the governing coalitions of center parties in several of these countries were short-lived, unable to agree upon policies capable of meeting pressing economic needs and to implement vigorously those measures upon which they could agree. Seriously threatened from within and seemingly able to do little more than maintain routine administration, the centrist coalitions gave the appearance of being caretaker governments that were sooner or later bound to be swept away by extremist movements of the right or left, or to collapse of their own factionalism and ineffectualness.

Reflecting and compounding the severity of these economic and political weaknesses was the conspicuous inability of the continental West European countries to make a significant contribution to their own defense during the alarming years of the developing cold war. None possessed the resources or the technology needed to make nuclear weapons, and even the raising and equipping of conventional forces were beyond their capabilities. The insurrection of the communists in Greece and their seizure of power in Czechoslovakia, the Berlin blockade and other initiatives and responses by the Soviet Union, and—most important of all—the outbreak of the Korean War in 1950 engendered a pervasive sense that Western Europe was in imminent danger of becoming the nuclear battleground of the third world war, which it was powerless to prevent and in which it would be incapable of defending itself.

As characterized in an analysis written at the time, the effects on West European morale of these postwar difficulties and crises, superimposed upon the retrospective sense of European failure, amounted to

a conviction—not always clearly articulated but felt nonetheless strongly—that the national political and economic structure of the continent is simply not adequate to cope with the rigorous world environment of the mid-twentieth century. . . . The average continental European feels himself a member of an enfeebled nation, the nearly helpless prize in a world power struggle in which his government plays no effective part. He knows that his economic horizons, his freedom of movement and opportunity are constricted within narrow national boundaries. He believes that the major factors determining his economic well-being, his military security and even his personal survival are beyond the capacity of his government to control or even to influence very much. Unlike the average American or Briton, he feels that his national state is no longer capable of adequately discharging the increasingly heavy responsibilities of political sovereignty. As a consequence, and no matter how much the traditions and culture of his society still mean to him, his belief in and loyalty to his government as a sovereign political entity, his willingness to sacrifice and, if necessary, to die for it have been very severely impaired.[3]

This widespread sense of the failure of the nation-state was reinforced by the conviction that European nationalism, the major cause of past wars, had to be superseded or securely constrained if world peace was to prevail. In the context of recent history, this meant essentially an enduring reconciliation of France and Germany. And, in the light of cold-war problems then developing, there was growing concern on both sides of the Atlantic that Germany had to be firmly tied to Western Europe lest it sooner or later come under Soviet control.

In view of the profound changes that have occurred in the nature of world politics since the early 1950s, it is difficult today to recapture the strength of Americans' concern lest the traditional European source of international conflict continue to constitute the major threat to the stability and peace of the new system of world order that the United States was seeking to construct in the postwar period. Of more immediate importance was the apparent need to prevent communist takeovers, particularly in France and Italy. And, to these defensive motivations, was added the growing conviction among Americans involved in U.S. policy making for Western Europe that the latter's internal problems could no longer be dealt with adequately by its small weak nation-states.

Conceptions of European and Atlantic Restructuring

The ideas and arrangements proposed on both sides of the Atlantic during the late 1940s and early 1950s can be divided into two kinds: those aimed at the eventual unification of Western Europe, and those for restructuring the relationships between a united Europe and North America, that is, for the organization of what soon came to be called the "Atlantic Community."[4] Each of these concepts embraced a range that varied with respect to the kind and extent of the economic integration and political and military unification envisaged and the nature of the relationship between the two parts of the region that would result therefrom. Moreover, each set was in varying degree both complementary to and incompatible with the other, depending upon the extent of the European or Atlantic unification believed to be required.

The predominant movement in continental Western Europe was inspired by the range of concepts envisioning as its maximum development a United States of Europe—a full federal union. Serious discussions of this possibility were carried on during World War II both in the resistance movements in the occupied countries and among people associated with the continental governments-in-exile. In the wake of the liberation, several private organizations were founded to promote various ways of achieving a united Europe. They soon polarized into the alternatives of the *functional approach,* explained below, and the *constitutional approach,* which envisaged the immediate calling of a convention to adopt a constitution for a federal union. In 1948, the alternatives were conceptually and organizationally consolidated in the European Movement, whose branches in the various European countries are still active today. Finally, in 1955, the Action Committee for the United States of Europe was formed as a multinational organization bringing together designated representatives of the major (noncommunist) European political parties and trade unions under the chairmanship of Jean Monnet.

The theory of functional integration essentially argues that, as the extent of economic integration among a group of countries increases, the concomitant need and pressure develop for supranational authority.[5] The maintenance of the economic integration already achieved and the management of further progress toward complete economic unification require increasingly close and continuous coordination of national economic policies, the settlement of disputes among the participants, and the formulation and implementation of joint measures to take care of common problems. As the pressure on them to carry out these important functions grows, the participating governments would be less and less able to agree upon and to implement effectively the necessary policies through negotiation and cooperation. Hence, they would have to delegate more and more of these responsibilities to nonpolitical, technically qualified agencies at the supranational level. The longer the integration process continues, the greater the power that would have to be given to superordinate authorities, who would thereby acquire more and more political, as well as economic, functions. At a certain point, their growing exercise of supranational power would be formalized through the adoption of a constitution for a federal union.

In this way, the functional and the constitutional approaches to union were reconciled. Although there have been some strategists who have continued to advocate, or have in recent years reverted to, the original constitutional approach, the predominant view in Western Europe has been that functional integration would eventually and inevitably lead to constitutional union.

The other set of concepts developed in the late 1940s and the 1950s dealt with the restructuring of relationships among Atlantic countries. They ranged from the liberalization of trade and capital flows, at one end, to the formation of an Atlantic union, at the other. Interest in the latter possibility was largely stimulated by the desire to find an Atlantic-wide alternative to European union. During the 1950s, its period of greatest significance, the Atlantic union movement was predominantly a North American phenomenon with considerable support in the U.S. Congress and American and Canadian business circles and with some adherents in Europe, especially in the United Kingdom. It, too, was envisaged as developing through either functional or constitutional approaches. The majority of Atlanticists had in mind a gradual functional approach, foreseeing the most likely course of evolution as occurring in NATO through the progressive unification of the armed forces and command structures of the member countries. Military unification would inevitably require close and continuous coordination of foreign policies, on the one hand, and of defense research and production programs, and hence of national economic policies, on the other. Either or both of these processes would lead to the establishment and strengthening of supranational authority which, in turn, would culminate in a formal political union in accordance with the theory of functional-constitutional inevitability. Except by the orthodox constitutionalists, who continued to urge the immediate calling of an Atlantic contitutional convention, the process of Atlantic unification was envisaged by its supporters as of much longer duration than the comparatively brief period of development which was all that the advocates of European union believed would be necessary to achieve their goal.

However, most Americans and Europeans concerned with the future development of the Atlantic region envisaged a less far-reaching restructuring of Atlantic relationships that would be consistent rather than competitive with European unification. This was the idea of

Atlantic partnership, an arrangement under which economic policy, foreign policy and defense policy for the Atlantic countries would be made jointly by the United States and a united Europe—the latter by reason of its unification able and willing to provide an equitable share. of the resources required to carry on the common Atlantic role in the world. Thus, European union was regarded as an essential precondition for the larger process of Atlantic integration. However, there have been important bodies of opinion on both sides of the Atlantic that have rejected this reconciliation—in Europe because it implied the eventual merging of a European union in Atlantic arrangements which, it was feared, would be dominated by the Americans; and in the United States because it was believed that the formation of a European union would eliminate the need for and the willingness of Europeans to participate in an Atlantic arrangement.[6]

Official policies in the United States and Western Europe were soon influenced by these postwar concepts and movements, although they tended to lag behind the private initiatives both chronologically and in the extent and pace of unification envisaged.

In the United States, official policy was dominated during the initial postwar years by the "One World" rationalism analyzed in Chapter III. This conception of a worldwide system of peace-loving independent states, large and small, willingly cooperating with one another directly and through a vigorous and increasingly effective United Nations, was the then current expression of the Wilsonian dream of a rational world order, under whose influence most of the senior government officials responsible for foreign policy had been educated and trained. To them, proposals for regional arrangements either in Europe or in the Atlantic area as a whole were not simply unnecessary but positively harmful because they were contrary to the globalist objective.

During the formative years of the European and Atlantic regional conceptions in the late 1940s, the advocates of these approaches within the U.S. government—mostly younger men educated and trained during the depression of the 1930s when Wilsonian rationalism was in discredit—were a minority engaged in an uphill fight. But events favored their cause. First, universalist expectations were more and more frustrated by the emergence of the global cold-war confrontation of the United States and the Soviet Union, and by the increasingly evident inability of the United Nations to bring about agreements among its sovereign member states on actions adversely affecting their national interests. Second, at both popular and official levels, European interest in European unification was growing rapidly during those years, and senior American officials found it difficult to resist the pressure for U.S. support of a regional approach both from their own younger subordinates and from their counterparts in the continental governments.

Official American policy was of crucial importance not only for Atlantic arrangements, in which the United States would participate, but also for European union, in which it would not. During the formative years of the late 1940s and early 1950s, U.S. influence was at its height and the U.S. will was as nearly unquestioned in Western Europe as it has ever been. The continental countries were so weak and dependent on the protection and assistance of the United States that official American opposition to the unification, and even to the economic integration, of Western Europe would have been sufficient to prevent either development. Conversely, the positive support and encouragement of the United States was a necessary precondition for their accomplishment. Indeed, an additional manifestation of the inadequacy of European nation-states was the fact that none of the continental countries possessed the will and ability to provide the requisite leadership toward European unification. Until the late 1950s, the policy of the United States was determinative, and the conflict of opinion within the U.S. government was of critical significance.

Two developments necessitated by external events proved decisive for this internal bureaucratic struggle. The first was the start of the Marshall Plan in the spring of 1948 and the establishment of a new U.S. government agency—the Economic Cooperation Administration (ECA)—to allocate and supervise the use of the immense funds appropriated for European aid. The ECA possessed not only the influence of money but also the intellectual power of a dedicated staff of younger professionally educated people drawn from the universities, business firms and other government agencies to temporary service in this challenging effort. In addition, it had the bureaucratic advantage of direct contacts with European political leaders and senior civil servants through its own missions abroad. Convinced by its analyses and the views of Europeans, the ECA was able to overcome the opposition of the State Department's universalists and to commit the U.S. government to encouragement of European economic integration, although it was initially unable to obtain open official support for European union.

The second development, promoted perforce by the universalists themselves, was the establishment of NATO in the spring of 1949 as an explicitly regional arrangement differing essentially from preceding peacetime defense alliances in the extent of the integration envisaged in command structures; in air, sea and ground forces; and in the standardization and production of armaments. However, the United States did not go beyond this Atlantic military integration until the Eisenhower Administration took office in 1953, when the President himself and Secretary of State John Foster Dulles openly and actively committed the United States to the support of European economic and political unification. Finally, in the early 1960s, top officials of the Kennedy Administration proclaimed as the "Grand Design" of U.S. foreign policy the eventual formation of an Atlantic partnership, which would be the consummation of the postwar effort to realize the twin goals of a full European union and a common Atlantic role in the world political and economic systems.

In continental Western Europe, too, a period of contending opinions was necessary before official policies were committed first to European economic integration in the late 1940s and then to European unification in the mid-1950s. Of decisive importance was the fact that the older politicians and officials of the interwar years, discredited by the failures of the depression and the Nazi conquest, were replaced in the immediate postwar period by a new generation of political leaders and civil servants with a strong sense of the inadequacy of the nation-state and a corresponding dedication to the unification of Europe. Robert Schuman and René Pleven in France, Konrad Adenauer and Walter Hallstein in Germany, Alcide de Gasperi in Italy, Paul-Henri Spaak and Paul van Zeeland in Belgium, J. W. Beyen in the Netherlands are only a few of the better known political leaders who became committed to European unification. In the late 1940s and the 1950s, they worked effectively with one another, with the growing group of Europeanists, headed by Jean Monnet, and with Americans in launching their countries on the unification process. In consequence of these efforts, Belgium, France, Germany, Italy, Luxembourg and the Netherlands became the founding members of the successive institutional arrangements that were intended to evolve inevitably into the United States of Europe.

In contrast to these continental countries, the United Kingdom emerged from World War II with the high morale and confidence in the future to be expected in a victor of that conflict. True, in the late 1940s and early 1950s, Britain faced reconstruction tasks and economic recovery difficulties fully as great as those of the continental countries, but its political system was unimpaired. Also, during this period, it was still able to contribute to the exchange of nuclear technology with the United States, and its own military establishment was still large and effective enough to sustain the conviction that it could continue to play a significant role in its own defense and in that of Commonwealth countries and client states throughout the world. Thus, despite their economic problems, the British had a strong sense of the adequacy of *their* nation-state and felt little, if any, need to join with the continental countries in the movement toward a political and economic union. The great majority of the British people in all social groups were convinced that they still possessed the strength and the obligation to play a major role in the world as the leader of a globe-encircling Commonwealth of nations and, through continuation of the "special relationship," as the closest and most influential ally of the United States. In the Scandinavian countries, too, the sense of inadequacy of their nation-states was not great enough to impel them to participate in the European unification movement.

Although the scope of European union was limited to the six continental countries, the initial institutional development from which the unification movement was later to grow embraced a much wider group of West European nations, as well as Greece and Turkey. This was the Organization for European Economic Cooperation (OEEC) established in Paris in the spring of 1948 to coordinate European participation in the Marshall Plan. The OEEC did not itself involve or directly lead to the transfer of sovereignty to supranational authorities. It did, however, make two essential contributions to the subsequent process of unification.

First, the OEEC firmly established in Western Europe the twin practices of mutual cooperation in the solution of common economic problems and of voluntary coordination of national economic policies among the member countries, and it worked out the conceptual techniques, procedural forms and staff functions required to make such activities effective. Second, the OEEC initiated and developed means for achieving a degree of economic integration among the West European countries that they had not known since the brief boom period of the late 1920s. Based on Marshall Plan aid, the OEEC was able to persuade its members to abolish the network of bilateral trade and barter relationships that developed after the onset of the great depression of the 1930s and was expanded in the immediate postwar years. Not only did these measures restore a multilateral system of trade and payments in Western Europe, but they demonstrated the effectiveness of institutionalized cooperation and policy coordination in achieving economic integration.

Progress of European Integration in the Postwar Period

The OEEC experience was a necessary prerequisite for the inception of the European unification process *per se,* which was initiated in 1951 by the six continental countries with the establishment of the European Coal and Steel Community (ECSC). This arrangement involved the abolition of tariffs and other restrictions affecting trade in coal and steel raw materials and products, and supervision of the current operations and planning for future growth of these industries by a High Authority exercising supranational regulatory powers. The favorable response to the ECSC proposal soon stimulated a more ambitious prescription for the Six to merge their armed forces and the control of their military production industries in a European Defense Community (EDC). However, despite strong pressure from the United States, the French Parliament refused in 1954 to ratify the EDC treaty, thereby aborting this aspect of unification—as well as the related proposal for a European Political Community (EPC) also envisaged in the treaty.

Although ardently pressed by the Europeanists, the projects for military and political unification were premature, directly and immediately requiring the surrender of central areas of national sovereignty. Hence, when the unification movement was resumed after this setback, it once more turned to the indirect strategy of functionalism. In a series of ministerial meetings during 1955 and 1956, the Six prepared the design for a common market covering not only coal and steel but all products traded among the member countries. These efforts eventuated in the Treaty of Rome, signed in 1957, establishing the European Economic Community (EEC) as of January 1, 1958. A separate treaty arranged for the creation of the European Atomic Energy Community (Euratom).

The Treaty of Rome provided in detail for the gradual formation of a customs union among the six member countries during a 12-year period terminating on January 1, 1970. Despite the difficulties encountered during the mid-1960s in the negotiation of a common agricultural policy—the precondition for free trade in agricultural products —and the problems posed by de Gaulle's opposition to supranationality, discussed in the next section, the EEC was able to accelerate to July 1, 1968 the achievement of its customs union. Moreover, in the same year, agreement was reached to merge the central agencies of the EEC, the ECSC and Euratom into the single European Community (EC).

The Treaty of Rome also envisaged that the customs union would in turn be only a transitional phase to a full economic union. It would gradually be achieved by adoption of common policies and regulations in all fields significant for intra-Community competition; abolition of the remaining barriers to the free movement not only of goods but also of capital, labor and enterprise; and development of a unified system of money and banking for the region as a whole. To foster this progressive unification movement, as well as to manage the process of forming and preserving the customs union, the Treaty established the European Commission endowed with certain supranational powers. However, the Treaty did not specify in detail—as it had for the customs union— the steps and timetable for achieving the full economic union or the political preconditions for and consequences of its attainment. The problems involved in the EC's moving beyond the customs union to full economic union and eventually to political federation, and the likelihood that these developments will occur are discussed later in this chapter.

Because economic and political unification was from the beginning the aim of the EC's founders, neither the United Kingdom nor the Scandinavian countries were willing to participate in its establishment. Nevertheless, they recognized the advantages of membership in a large free-trade arrangement and feared the possible adverse consequences for their own exports of the trade-diverting effects of the formation of the European Community. To obtain the benefits of economic integration without supranationality and unification objectives, the British initially proposed the formation of a European-wide free-trade area to include the Community. When this proposal was rejected by the Six as an effort to sabotage the Community, the United Kingdom joined with Austria, Denmark, Norway, Portugal, Sweden and Switzerland (with Finland and Iceland participating later) to form the European Free Trade Association (EFTA) in 1960. The EFTA obtained a waiver from the General Agreement on Tariffs and Trade (GATT) permitting its members to retain trade restrictions on most agricultural products, thereby avoiding the difficulties faced in this area by the European Community and making possible the achievement of internal freedom of trade affecting industrial products by January 1, 1967.[7]

The EFTA's progress during the 1960s was all the more noteworthy because its existence was in doubt throughout the period of its successful movement to free trade in industrial products. Although initially proposing the EFTA as an alternative to a supranational arrangement, the United Kingdom, the leading member, soon reversed its policy toward European unification. In 1961, it started negotiations regarding membership in the Community, and the British example was immediately followed by two other EFTA members. This change in British policy reflected not only the political motivations noted in Chapter IV but also the United Kingdom's increasingly serious economic problems in the early 1960s. In part, too, the decisions of the United Kingdom and other EFTA countries to seek EC membership were influenced by the sense of progress and the growing prestige that characterized the Community in those years.

For, in the early 1960s, the EC's six members were enjoying unusually rapid rates of economic growth, expanding trade, rising living standards, full—indeed, overfull—employment, increasing monetary reserves, and a pervasive feeling of economic well-being and continuing momentum. Their joint efforts to meet the schedule for establishing the customs union specified in the Treaty of Rome and to work out the policy measures required for it were being conducted in the "Community spirit," as it was called, of willingness to subordinate national interests to the new interest of the common objective, a united Europe. And, conflicts of national interests that in other circumstances would have been irreconcilable were in fact settled in the spirit of community. In turn, these successes further strengthened the sense of progress and the conviction—not only among the Six but also in the other Atlantic nations—that the Community was advancing rapidly in the unification process, which would irresistibly bring it to full economic and political union in the foreseeable future.

So great was the EC's self-confidence and *élan* in those years and so high its prestige that the attitudes toward it of the other Atlantic countries were correspondingly affected. Not only did the United Kingdom and other European nations begin negotiations for membership in the Community, but, equally significant, the United States regarded it with mounting respect, and even with some concern. Among official policy makers in Washington and opinion leaders throughout the country, there was growing agreement that the United States would have to adapt its economic, political and military relationships to the new capabilities and challenging potentials of a united Europe. It was this reaction in the United States that led in 1962 to the proclamation of the "Grand Design" for Atlantic partnership, under which the U.S. government offered—verbally, at any rate—to share equally with a united Europe in the responsibilities and costs of managing the security and progress of the "Free World." Thus, in the early years of the 1960s, it seemed that both the Europeans' goal of union and the Americans' goal of partnership were at long last within reach.

European Unification in the Transition to the Period of the New Nationalism

Yet even midst the self-congratulations and confident predictions of official spokesmen and opinion leaders on both sides of the Atlantic in the early 1960s, the developments that were in the course of that decade to bring to a halt the movement toward European union—and to frustrate American expectations of Atlantic partnership—were already beginning to manifest themselves. In the ensuing years, it became customary to attribute the causes of these trends, as well as the manner of their expression, to General de Gaulle, and to insist that the unification process would surely be resumed after his departure from political office. Although understandable in the circumstances both of de Gaulle's provocative arrogance and of the emotional intensity of the dedicated Europeanists, these views reflected the persistence of postwar expectations for the future rather than recognition of the emerging realities of the new period in European and Atlantic relationships then in its transitional phase.

Certainly, de Gaulle's political style and much of the specific content of his conception of the roles that France should play in Europe and Europe in the world were personal characteristics unlikely to be found in his successors, even in those that might have charismatic authoritarian personalities. However, neither de Gaulle's manner nor his nostalgic efforts to revive something like the mid-19th-century system —with France rather than England holding the world balance of power as the leader of a *Europe des patries*—were responsible for the altered prospects of the postwar goals for transforming the Atlantic region. At most, they were adventitious factors that hastened the emergence and helped to articulate the expression of fundamental changes in the three major systems—national, regional and worldwide—whose interactions determined the limits within which and the directions toward which policy choices were made by Europeans and Americans from the mid-1960s on.[8]

These developments in the three systems during the 1960s gradually altered European feelings regarding the adequacy of the nation-state. Although changes in the sense of adequacy of the nation-state cannot be measured precisely, they can be assessed qualitatively in several ways. Perhaps the simplest and clearest method is to compare the situation that emerged in the period of the new nationalism with that of the earlier postwar years.

In contrast to the then pervasive fear of imminent Soviet invasion, fewer and fewer West Europeans were preoccupied in the course of the 1960s with worries about an impending external aggression. Even the two apparently most directly threatening Soviet moves in the decade —the Berlin Wall crisis of 1961 and the occupation of Czechoslovakia in 1968—were in time seen in Western Europe to be defensive actions designed in large part to counteract the attraction it exerted on the East European members of the Russian hegemony. True, suspicion of Russian intentions and apprehension about Soviet military capabilities continued to be widespread—indeed, they have been felt even by some West European Communist Party members and sympathizers. But, these concerns have become far less intense and pressing than they were in the late 1940s and early 1950s, when the Soviet threat seemed at its height. In the course of the 1960s, it appeared more and more improbable that the Soviet Union would switch to an expansionist strategy in Western Europe in the absence of changes in American policy that would encourage such behavior.

Thus, insofar as fear of the Soviet Union persists, the American nuclear guarantee has been—and is likely to continue for some time to be—sufficiently credible to provide an offsetting reassurance. Nor has another external menace appeared to take the place of the Soviet Union. Efforts to cast the United States in this role have very little credence, and the possibility that China may become so in the future, although acknowledged, has virtually no present effectiveness in Western Europe. The absence of an external threat that appears so ominous as to confront Europeans with the need to "unite or perish" probably contributed more to restoring faith in the adequacy of the nation-state than any other single factor.

This judgment is not meant to belittle the importance of the positive elements involved. Certainly, the changes in political conditions in Western Europe since the early postwar years have enhanced the acceptability of the nation-state. Except in Italy, West European governments ceased to be regarded as impotent "caretakers." This was true even though domestic political problems persisted in some countries, notably France and Italy, and may soon arise in others, such as Spain and Portugal. In general, these difficulties are manifestations of the slowness of political and administrative institutions to adapt to basic changes in attitudes and social relationships, of disagreements over resource allocations, and of the persistence of the older types of class antagonisms. Except for the single-minded Europeanists, no significant body of opinion in the countries concerned believes that these kinds of political problems could be eliminated, or even substantially mitigated, by the transfer of sovereignty from national to supranational institutions.

The major positive reason, however, for the renewed sense of adequacy of European nation-states has been their extraordinary success in adopting and carrying out the policies and programs that have contributed so importantly to the region's unprecedented economic growth and rising prosperity. Despite the serious and persistent problems of internal and external imbalance, the economic conditions and prospects of the West European countries have become the opposite of what they appeared to be in the postwar years. This economic reversal is much more complete and dramatic than the political change. There is no question today of the economic viability of West European nations, and the circumstances in which they might again become dependent on American aid are hard to imagine. The economic uncertainty regarding European nation-states no longer relates to sheer economic survival, as in the 1940s, but to whether or not they can preserve and increase the prosperity already achieved without further progress toward European union, or continued Atlantic integration, or both.

Increasingly in the course of the 1960s, the answers to these questions were shaped by the dynamic tensions between the emerging pressures of the new nationalism and the growing constraints of European economic integration. On the one hand, European governments were perforce becoming more responsive to elite-group and popular demands for the attainment of a broadening range of competing national welfare objectives, and the resulting politicizing of the process of allocating resources had the effect of continually expanding governmental responsibilities and functions. On the other hand, their freedom of action to achieve national goals was being increasingly constricted by their mutual commitments to remove trade and payments barriers and to coordinate their national economic policies in the European Community and the EFTA, as well as to transfer small, but nevertheless significant, portions of their sovereign powers to the EC's supranational Commission. The resulting contradictions lie at the root of the difficulties that more and more impeded the European unification movement in the 1960s and will continue to confront it in the foreseeable future.

The major achievements of the European Community during the 1960s were the completion of its customs union ahead of schedule, the adoption and implementation of a common agricultural policy, the inception of negotiations regarding other common policies significantly affecting the conditions of competition (such as taxation, social-welfare benefits, transportation and energy costs, business regulations, product quality and safety standards, and government procurement practices), and the conclusion of a growing number of preferential trade and investment agreements with nonmember countries in Southern Europe, the Mediterranean and Africa. These advances, especially the first two, are noteworthy not only for their result—the forging of a single "common market" from six separate national markets—but also because of the magnitude of the obstacles that had to be overcome. Indeed, it was these difficulties increasingly generated by the new nationalism, as well as by General de Gaulle's contrary views, that were responsible for the two major failures of the European Community during the 1960s—the French veto in 1963, and by implication again in 1967, of British membership; and the refusal to grant the EC's central institutions all of the supranational powers specified in the Treaty of Rome, let alone to extend them beyond those limits.

The resistances arising from the new nationalism account for the fact that the advances of the 1960s did not go significantly beyond measures of economic integration. Although they are prerequisites for progress toward economic and political union, such steps are not in themselves of crucial importance for that objective. Hence, the future of the Community will be determined by whether in the course of the 1970s its members are willing to undertake the further institutional developments necessary to move it decisively "over the hump" toward full union.

These institutional changes must sooner or later involve the shift of certain essential economic and political powers from the national to the supranational level. For decisive unification, the central authorities will need *at a minimum* (a) the economic power to levy and dispense significantly large taxes, control the money supply, regulate interest rates and credit availabilities, and manage external monetary and commercial relations; and (b) the political power to carry on foreign relations and control the major military forces. Whether the supranational agencies are democratically elected or appointed by member governments is not a critical constitutional issue, although in practice it may be important in determining whether the central authorities will be sufficiently independent of national governments.

Early in 1971, the EC's member governments agreed upon a new approach designed to initiate such a process of institutional transformation. This was the project for an economic and monetary union, that is, for the eventual establishment of a common currency and the transfer to the EC's central authorities of the requisite monetary and fiscal powers. Suspended during the prolonged dollar crisis of 1971, this project was reactivated in the spring of 1972.

The development of a monetary union is supposed to occur by stages. The first phase involves narrowing the margins for fluctuations in the rates of exchange among members' currencies, and more effectively coordinating their economic policies through regular Community reviews of national budgets and central bank measures prior to their adoption. Commitments were also made to provide short- and medium-term credits to member states with insufficient reserves to keep their exchange rates within the narrower margins. While the reduction of exchange-rate fluctuations and the provision of credits are mandatory, the coordination of national economic policies is still voluntary. During the first stage, national administrations and central banks are not required to adopt the advice they receive from one another and from the Commission in the periodic reviews of their budgets and other *macro*-management policies. However, the second stage would involve elimination of all fluctuations among the members' exchange rates and the grant to the Community's central institutions of the power to enforce, if necessary, the required coordination of national economic policies and conditions. Moreover, unless such measures are taken within a specified period—initially fixed as not later than January 1, 1976—member governments would be released from the mandatory exchange-rate and credit commitments of the first phase. Thus, the critical decisions as to whether the Community will simply continue to deepen its degree of economic integration on the basis of voluntary cooperation or will begin to adopt the crucial supranational measures requisite for unification are not likely to be made before the second half of the 1970s.

During the intervening years, the Community is also confronted with the equally formidable task of absorbing new member states and negotiating association agreements with additional countries in Europe and Africa. Successful in 1971 in negotiating membership in the Community, the United Kingdom, Denmark, Ireland and Norway have since been engaged in adapting themselves to the existing EC arrangements. For the remaining EFTA countries—Austria, Finland, Iceland, Portugal, Sweden and Switzerland—it was important to preserve their mutual free-trade arrangement, to continue to enjoy freedom of trade with the United Kingdom, Denmark and Norway, and to obtain free access to the Community as a whole. Hence, they have been working out association agreements with the Community that assure these benefits in greater or lesser degree. Not only does the enlargement of the Community entail difficult changes but it also significantly affects the EC's willingness and ability to undertake the supranational developments required for economic and monetary union.

Popular and Elite-Group Attitudes Toward European Unification[9]

An assessment of the probability that in the course of the 1970s Europe would adopt such crucial institutional advances has to take into account the psychocultural, as well as the economic and political, elements that help to shape the choices that Europeans generally and the elite groups in particular will make in the years ahead. The former are analyzed here; the latter in the next section.

The great majority of the population in West European countries, including the United Kingdom, is intent upon the achievement of those national goals primarily affecting their economic and cultural welfare. When the people generally become aware of the competition among national goals, they tend to resent the allocation of resources to those that demonstrably interfere with the maintenance and improvement of their living standards. This reaction is especially marked with respect to objectives and responsibilities beyond their borders that appear to make only small or deferred contributions to domestic welfare. Thus, in the course of the 1960s, the majority of the people increasingly looked inward to their national political and administrative institutions— rather than outward either to American aid, as in the 1940s and early 1950s, or to new supranational authorities, as in the late 1950s and early 1960s—for preservation and continued improvement of their economic and cultural welfare.

Although the popular sense of national identity has been strengthened in this way, the developments of the postwar period and the 1960s have nevertheless made many people more conscious of a parallel European identity. The passing of European empires and the rise and conspicuousness of many new nations with non-Western cultures; the disproportionate power and worldwide interests of the two superpowers; and, above all, the progress of European integration have fostered the sense of European identity *vis-à-vis* the rest of the world, including the non-European members of the Atlantic region. However, the commitment to a united Europe of people generally tends to be passive rather than active, protecting rather than aggrandizing. Hence, they are devoted to a united Europe and to their nation-state in a manner that does not regard these two loyalties as incompatible or even as noticeably competitive. And, because both commitments are passive, most people are not prepared to make major sacrifices in terms of their economic and cultural welfare either for their countries individually to play leading roles in world politics or for their governments to press on with the unification of Europe.[10]

Popular attitudes are important not because they directly determine European policies in a positive sense but because they set general limits to elite-group policy making and execution. It is significant, for example, that the Gaullist party could survive in France only without de Gaulle—that is, without the leader most strongly identified with the diversion of resources to equipping France for a proto-superpower role and with a foreign policy bound to alienate the United States. Certainly in the shorter term and perhaps in the longer term as well, other activist political leaders on the continent and in the United Kingdom who may try to press strongly either for an independent proto-superpower status for their countries or for European unification at the cost of domestic economic and cultural objectives would be likely to suffer a similar fate.

The attitudes of the elite groups are more varied and complex. Broadly speaking, three strands of opinion can be distinguished. But, the differences are not clear-cut and individuals in one category often hold some opinions representative of another category even though they may be partly or wholly contradictory.

The first category consists of the Europeanists. Many of the politicians and opinion leaders responsible for organizing and guiding the European union movement during the postwar period died or retired from public office in the course of the 1960s. Although some of them continue to provide leadership to the unification effort, their influence naturally declined after they ceased to be active in political life. Nor does the newer generation of political leaders, who have been taking their place, possess as much stature in national politics as their predecessors did in the immediate postwar years. The latter had the prestige both of their own wartime resistance records and of active support by the victorious Americans during the years of most strongly felt European dependence on the United States. In contrast, the upcoming generation of Europeanist politicians are not replacing discredited elders; quite the contrary, they are continuing to serve the objectives and to follow the policies formulated by their better-known and more prestigious predecessors.

A sense of Europe's redemptive mission in the world characterizes both the older and the younger generations of Europeanist leaders, but with certain important differences. Both share the conviction, validated by the history of Western civilization, that European society and culture were in recent centuries the major sources of creative advances in the socioeconomic welfare and cultural enrichment of mankind; and they are convinced that, once united, Europe would continue to make great contributions in all fields of human endeavor. Moreover, they believe that European experience and wisdom are essential for tempering the dangerous rivalry of the superpowers, restraining the rashness and impatience of American redemptive activism, and guiding the new nations of Asia and Africa toward orderly and rational behavior at home and abroad. But, whereas the older generation of Europeanists, and especially the great names of the movement, envisaged Europe's mission as being fulfilled in close harmonious partnership with the United States, the younger generation is inclined to think of a united Europe as an independent active power—indeed, a new superpower—in the world.

It is significant in this regard that, with the decline of the Soviet menace as a compelling spur to European unification, some of the younger Europeanists on the continent and in the United Kingdom have been trying to substitute for it the American menace. This is not the danger of possible conquest by the United States, which would not be widely credible in Western Europe, but the much subtler and largely unintentional threat of being reduced to economic satellites of the United States in consequence of the "technological gap," the spread of giant American corporations, and the monetary supremacy of the dollar. Others stress the positive implication of the technological gap for European unification—that, only with nuclear, electronic, computer and aerospace research capacities and productive facilities equal to those of the United States, would a united Europe have the industrial and military power for playing an active, independent and directive role in world politics. It is by no means coincidental that the branches of industry in which European publicists claim that the technological gap is most significant and needs most urgently to be closed are precisely those that are of greatest importance for the achievement of the nuclear capabilities required for superpower status.

The second category comprises the technocrats, especially the large, upcoming age groups of civil servants, business managers, natural scientists and engineers, economists and sociologists, and others, trained to use specialized skills and sophisticated techniques in their work. In the decades since World War II, the various kinds of technocrats have been becoming steadily more numerous and influential in the policy-making levels of government ministries, business firms, educational and research institutions, professional associations, scientific societies, and other organizations requiring technical and administrative capabilities. In the course of the 1960s, they began to supersede the older, much less technically and professionally trained generation in top leadership positions, and will increasingly do so during the 1970s. By reason of their numbers and of the economic and political importance of the institutions which they dominate, the range of attitudes and opinions characteristic of the technocratic elites is of considerably greater significance than that of the Europeanists *per se*.

The majority of technocrats are in favor of European unification because its benefits are believed to outweigh its costs—in other words, because it is a rational goal. Their commitment to a united Europe, is, therefore, a rational one, by and large lacking the passionate conviction of necessity that constitutes the driving emotional force in the ardent Europeanists. Moreover, to the technocrats, European union is only one of the many changes needed to bring about the rational reorganization of society, which is their vocational purpose as a major social group. (To the Europeanists, union is the quintessential precondition for all the rest.) The technocrats tend to believe that, because rational considerations are working for the uniting of Europe, it is bound to be achieved sooner or later. Their conception of the unification process is as a kind of passive functionalism, in contrast to the actively probing and pushing functionalist tactics of the Europeanists.

In effect, the majority of technocrats favorable to European union take it for granted. As individuals, their attention and energies are focused primarily on applying their knowledge and skills, through the public and private institutions in which they work, to advancing the personal, organizational and national interests involved. Those who are motivated by emotional commitments beyond purely self-interested goals are concerned with modernizing and increasing the efficiency of particular organizations and institutional systems (such as the economic, governmental and educational systems) or of the society as a whole. The resources needed for these purposes are produced and allocated through national economic, political and administrative processes. The political power and institutional means that the technocrats require for effectuating the necessary policies and programs are mainly provided by the ministries and other public agencies of their nation-states. In consequence, both their most strongly felt concerns and their most significant activities tend to be inward oriented. Nor are they likely to look toward the central institutions of the Community and to flock to Brussels to work in them unless and until its supranational authorities acquire resources and powers of policymaking and implementation more comparable to those of national governments. The technocrats will climb on the bandwagon of European union only when it is moving decisively toward its goal.[11]

The technocrats' inward focus of concern is, however, more qualified and contingent than that of the people generally. Their positivism and their pride in the historical achievements of Europe and of their own nations impel them to an activist redemptive conception of their individual and social responsibilities. Thus, they are not fundamentally opposed to important international roles for their countries; they are not isolationists as a matter of principle. Rather, they recognize that neither their nations individually nor a united Europe can achieve active, independent superpower status unless and until the process of internal transformation has gone much further than it is likely to do in the shorter term. Hence, the current disagreements between the technocrats and the people generally are not about their respective assessments of the relative importance of domestic welfare goals compared with international objectives. Their principal differences are over the relative priorities of different competing *domestic* claims on resources —as, private consumption versus public investment—and over popular objections to the domestic consequences of many of the measures by which the technocrats try to deal with internal and external imbalances —such as monetary and fiscal restraints and incomes policy.[12] The majority of technocrats tend to believe that, for the 1970s at least, internal goals must continue to have a prior claim on attention and resources, regardless of whether or not the priorities between domestic and international objectives would be reversed thereafter. For example, efforts like General de Gaulle's to force the pace by diverting resources from institutional improvements to military and foreign-policy purposes were criticized by most French technocrats as premature and ill-advised, even though they were pleased at the time by France's resulting greater importance in world affairs.

There is, however, a minority of technocrats whose attitudes include important exceptions to the foregoing characterization of the majority. While they, too, are concerned to achieve internal modernization and welfare goals, their activism is much more outwardly oriented, and they are, in consequence, prepared now to limit resource allocations to domestic purposes in order rapidly to develop the means for enabling their countries to play independent and important roles in world affairs. Members of this minority include the French technocrats who were committed by conviction as well as by interest to the policies and aspirations of General de Gaulle, and their counterparts in other European countries, especially Germany and the United Kingdom. Some of these more chauvinistic technocrats, such as the orthodox Gaullists, were and continue to be strongly opposed to European union; others favor it on the assumption that their own country would surely dominate it.[13] And, both types tend to be overtly anti-American, as do the remnants of the older pre-World War II kind of aggressive xenophobic nationalists.

Because of the nature of long-term sociocultural trends in Atlantic societies, whose future development is projected in the final chapter, the technocrats will increasingly comprise the most powerful opinion-forming and policy-making groups in the Atlantic countries. There are, of course, numerous and often serious conflicts among them that express competing institutional and individual interests, partisan and personal rivalries, and differing perceptions of particular problems and divergent conceptions of their solutions. Nonetheless, the basic similarities in their ways of seeing, feeling, thinking and acting make for a broad consensus on national goals and priorities, on methods for dealing with the difficulties of maintaining internal and external balance, on the costs and benefits of European union and Atlantic integration, and on the advantages and disadvantages of existing and alternative relationships with the United States. Just as popular attitudes set the general limits within which political leaders can act, so the ideas, expectations and prescriptions of the technocrats in government ministries, business firms, educational and research institutions, and so forth, provide most of the specific content of actual national policies and programs, as distinct from the rhetoric of partisan politics. Politicians tend more and more to legislate and administer in accordance with the attitudes, ideas and methods of the relevant technocratic groups. In many cases, too, especially among younger people, politicians are becoming more and more technocratically oriented through general education and the influence of prevailing elite-group opinions, and some are themselves technocrats by formal professional training and occupation.

The third category of opinion-forming elites consists of dissenting students and other young people, along with their older intellectual mentors and followers in literary and philosophical circles. Their attitudes cover a wide spectrum of general dissatisfaction with existing Atlantic societies and cultures, ranging from individualistic idealism to the more disciplined revolutionary activism of the New Left and the Edenic spontaneity and philadelphic communalism of the neoanarchists. Often competing furiously among themselves, the different forms and factions of the student and youth movements are important for national policies not because their views are likely to become predominant in the future but because of their present influence on popular and elite-group opinions in their countries. Active student and youth commitment to particular national goals, domestic or foreign, has in the course of the 20th century become an increasingly significant—although by no means the determinative—factor tending to raise the priority of these aims and to augment the attention and resources devoted to them. Conversely, by depriving certain goals of such support, the opposition or indifference of the students and articulate young people weakens the efforts of other groups in the society to achieve these objectives.

In Europe as in North America, a minority of students and other young people dissent from what they believe to be the injustice, exploitation, dehumanization, mechanism and philistinism intrinsic in the fundamental constitution of the rationalized, technocratic societies and positivistic cultures of the Atlantic countries. Their opposition is not specifically directed against the existence of separate national sovereignties, nor do they look to European political and economic union as the remedy for the evils they decry. Rather, their aim is to transform the nature of institutions and values *per se,* not to restructure the way they are organized from the independent nation-state form into the federal Europe form.

Similarly, except for the Communists and some of the New-Left groups, European student and other young dissenters are intent upon the renovation of their own societies and cultures rather than upon fulfilling a sense of mission to help transform those in Asia, Africa and Latin America. Nor are most dissidents very much more hostile to the United States than they are to the institutions, values and persons that exemplify the evils of their own societies and cultures. True, they demonstrate against U.S. intervention in Vietnam and condemn American imperialism and exploitation. Nonetheless, their anti-Americanism tends less to express a strongly felt concern for the welfare of Asians or Latin Americans than to reflect their belief that American society and culture are the paradigms of some—though not all— of the institutions and values they hate most in their own countries.

In essence, most students and other young people are postnational, not antinational. They already have a sense of being European, as well as British, French or German. Moreover, these loyalties are largely cultural rather than political, and they are generally accepted as facts of life requiring neither passionate affirmation nor protective defense. Hence, they are not more strongly committed to unifying Europe than they are to perpetuating the independence of their own nation-states. Instead, they see themselves as fighting their battles on other terrain and for other purposes. In their own ways, therefore, they too fail to give much effective support to the Europeanist commitment to unify Europe, and very little, if any, to the superpower aspirations of the minority of nationalistic technocrats and of the remnants of the prewar chauvinists.

European Integration Versus Unification

Since the 1940s, the Europeanists have based their expectations of achieving economic and political union on the theory of functionalism. Nor can there be much doubt that, up to a point, functionalist theory is logically valid and empirically verifiable. The experience of economic integration both in Western Europe and in the Atlantic region as a whole demonstrates that this process initiates new and strengthens old pressures at both private and governmental levels for further progress in removing discriminatory policies and practices of all kinds and for greater harmonization of economic conditions among the participating countries. All other things being equal, the self-reinforcing character of functional integration would sooner or later bring about dynamic tensions among interests, pressures and problems that would constrain the transfer of crucial political and economic powers to central authorities. However, necessary as the *ceteris paribus* qualification is for theoretical analysis, it is rarely valid in real-life situations. Many other factors—economic, political and psychocultural—besides those involved in the self-reinforcing tendency of functional integration also exert powerful influences. The main factors nullifying or impeding this tendency are surveyed first, then those that reinforce or accelerate it.

Will Integration Necessarily Lead to Unification?

Functionalist strategy implicitly assumes that national governments and private interests have no choice other than to resolve contradictions and eliminate problems even if these results can be accomplished only by transferring responsibility for them to supranational agencies. That there is a decided preference, and hence a marked tendency, to remove conflicts and difficulties rather than to endure them is undeniable—else the human race would still be living in caves. But, there is no compelling necessity to do so. The conviction that all problems must and can be solved reflects the ethical imperative and rational faith of technocratic positivism, not the realities of human experience. It is never inevitable that logic and the problem-solving impulse will prevail, only more or less probable. History equally demonstrates that people can and do live indefinitely with contradictions and problems. They may lack the knowledge or the resources for solving them. The opposing feelings and interests may be in balance. The benefits that would be obtained or the difficulties that would be eliminated may not be sufficiently greater than the sacrifices involved to motivate the necessary actions. Thus, the nation-states comprised in the European Community may prefer to endure indefinitely some or all of the problems generated by the disparities in economic conditions and the divergences in economic trends among them rather than transfer additional important aspects of their sovereign power to supranational authorities.

Functionalist inevitability is sometimes asserted in the form of the "either/or" fallacy: either the EC's customs union will move ahead to full economic union or it will surely disintegrate into its original national components. But, often in life, as always in logic, one extreme or the other is not the only possibility. In the case of economic integration, the process can continue indefinitely or can even broaden and deepen without changing its essential character. That is, national governments can agree to limit the use of their sovereign powers without transferring them to a supranational authority; and the necessary degree of harmonization among their national economies can be accomplished by the freer operation of market forces guided, offset and supplemented by the voluntary coordination of national economic policies. True, disparities in the conditions of competition within this kind of an arrangement are likely to be greater and its means of coordination less efficient than they would be in an economic union directed by a supranational authority. Yet, *ceteris paribus,* neither drawback would generate such severe strains as to necessitate a choice between dissolution or unification.

The EFTA's experience of economic integration under voluntary self-restraint and cooperation is especially illuminating. The EFTA achieved freedom of trade in industrial products more rapidly than the

European Community. Moreover, it weathered the potentially disruptive crises of the British import surcharge in 1964 and the British devaluation in 1967 as successfully as did the Community its members' changes in exchange rates. More important, the EFTA made much greater progress than the Community during the 1960s in coordinating certain national disparities that significantly affect the conditions of competition, notably government procurement practices, product standards and testing, and restrictive business practices. And, it did so by voluntary agreements and not by supranational control. Specifically eschewing a unification objective, the EFTA dealt with the potentially divisive problems of economic integration on a pragmatic basis by means of voluntary coordination.[14]

Thus, continuing economic integration need not necessarily lead to unification. Voluntary coordination among national governments is neither too difficult nor too inefficient to cope with many critical issues and problems *when sufficient willingness to reach agreement exists.* And, when it does not, national governments are even less likely to grant to a supranational institution the power to impose solutions on them. Rather, they would be prepared to live indefinitely with the difficulties and deficiencies involved.

National politicians and civil servants can avoid, postpone or slow down negotiations for further development of the EC's central institutions and supranational authority because the pressures on them that functionalist theory assumes would inexorably work for increasing unification are not directed solely and cumulatively toward that end. In practice, both the national and the private interests involved are ambivalent—that is, the rational considerations at stake are not decisively on the side of unification. On the one hand, the benefits of further economic integration and the obligations they have undertaken to advance it impel national governments to consider seriously the proposals for common policies and for supranational developments made by the EC's Commission and to participate in the negotiations concerning them. Private business firms, too, recognize the advantages for them of the equalization of the conditions of competition within the Community that is the general aim of most of these initiatives. On the other hand, further economic unification would mean equivalent losses or limitations of authority by national governments over important aspects of their economic systems and social-welfare processes. Not only are national politicians and bureaucrats unwilling to "put themselves out of business" but they are reluctant to impair their ability to fulfill their basic responsibility for assuring national survival and well-being. Similarly, private interests are reluctant to forgo the benefits they have been deriving from differences in national conditions and policies, which generally have the effect of discriminating in their favor. The familiar bird in the hand is often believed to be worth more than the as yet unknown birds in the bush.

The Political Perplexities of Unification

These ambivalences of national and private interests are substantially magnified by the political and psychocultural elements also inherent in the process of further unification. For, the greater the authority that is acquired by the supranational institutions in Brussels, the more critical the political question of who will control them becomes. It is a tribute to the good sense of contemporary West Europeans, as well as a sign of the passing of the older form of xenophobic nationalism, that so little has been written or spoken on this crucial aspect of European unification. Nevertheless, the uncertainties involved are major considerations in the minds of many political and opinion leaders in Western Europe, as well as of the people generally.

So far, the issue has not arisen in a positive sense, although there have been recurrent complaints of excessive French influence. Nevertheless, the question of ultimate political control has been important negatively in inhibiting agreement on measures of further unification that might otherwise have been adopted. It has also been, and is likely to continue to be, one of the three main obstacles to the formation of a European nuclear force—the others being the unwillingness to divert substantial resources from nonmilitary national goals to this purpose and the persisting, albeit diminished, credibility of the U.S. nuclear guarantee. For, once the central institutions acquire military control over nuclear weapons and the economic power to tax and regulate money and credit, they would possess the external and internal essentials of political sovereignty. The constituent national society or elite group able to exercise the preponderant influence owing to its size, wealth, dynamism or skill would sooner or later dominate the emerging union.

The issue of political control is discussed by many Europeanists as though it were simply a matter of establishing at the proper time the necessary constitutional arrangements for some form of popular election of a European parliament and for supervision by it of the supranational executive agencies of the Community. They envisage that the political aspect of this change would involve the transfer of domestic politics to the European level—that is, the various national political parties would coalesce in accordance with their conservative, centrist or radical orientations, and European politics would thereafter consist of the same kinds of interest-group competition and bargaining, and disagreements over goals and resource allocations that now constitute much of the substance of national politics. And, it is probable that such a trend would develop, as presaged by the fact that, in the existing advisory European Parliament, the representatives sit in accordance with partisan, not national, affiliations.

However, this aspect of the process of political unification is already, and will continue to be, permeated, distorted and partly offset by another trend that reflects the momentum of national institutions, interests and senses of identity in the period of the new nationalism. This trend began to manifest itself in the mid-1960s—initially in the bitterness engendered by de Gaulle's veto of British membership in 1963; more strongly in the contentious and prolonged negotiations over the price provisions of the common agricultural policy in 1964; and fully in the so-called "crisis of 1965" over the financial arrangements for the common agricultural policy and the underlying issues of Commission versus national-government responsibilities and functions. One important casualty of these experiences was the "Community spirit" of subordinating national interests to the common purpose that had played so crucial a role in the EC's progress during its early years. It was customary to attribute not only the specific timing and mode of expression of this nationalizing trend but also its cause to General de Gaulle and to expect that the "Community spirit" would be revived after his departure from office. The fact that this trend has not vanished or even substantially diminished since de Gaulle's death would indicate that it is rooted not in the General's personality but in the attitudinal and institutional changes that characterize the new European nationalism.

Indeed, this trend toward nationalized politics will probably strengthen rather than diminish in the years to come because it is fostered both by the existing institutional arrangements of the Community and by the unlikelihood that a popularly elected European parliament with effective powers would be established soon enough for it to stimulate sufficiently the trend toward Europeanized politics. Since the mid-1960s, the Council of Ministers, consisting of member-government representatives, has increasingly asserted its influence over the Commission, the Community's embryonic supranational authority. In consequence, the latter has been playing less of a leadership and policy-making role and becoming more of a technical planning, implementing and advisory agency. This development facilitates the expression of national interests and bargaining power in the Council and the application of national pressures on the Commission and secretariat. In contrast, the existing European Parliament—even though it is composed of national representatives sitting in accordance with partisan affiliations—has no legislative powers and can only review the work of the Commission and make recommendations to the Council. While it could expose and deplore a growing exercise of national influence, it lacks authority that would permit it to counterbalance, if not to arrest, the trend toward nationalized politics. And, as that trend strengthens, it would make less and less likely the granting of effective powers to the Parliament.

The expansion of the EC's original membership to include the United Kingdom, Denmark, Ireland and Norway also inhibits the development of the trend toward Europeanized politics. These countries did not experience in the postwar period a sense of the inadequacy of the nation-state comparable to that of the EC's founding members. Nor did the serious economic problems of the United Kingdom in the late 1960s and early '70s generate such feelings among the British people. Despite the professed commitment of many British elites to European political and economic union, it is more probable that the United Kingdom—like France—will try to slow down the unification process rather than concur in, much less push for, the transfer of sovereign powers to supranational agencies.

It is hard to believe that three such identity-conscious, former imperial powers as France, Germany and the United Kingdom have so lost their sense of vocational mission and conviction of superiority that their younger activistic elite groups would refrain in the years to come from trying to use their size, prestige, economic power and organizational skills to compete for the leadership position in an emerging union. Yet, neither the British nor the French nor the Germans would be willing to participate in a European union dominated by one of them. Nor, for all their genuine devotion to the unification goal, is it likely that the smaller European countries would continue to press for its achievement if such rivalry of the big three for preponderant influence were to become evident. While they might reluctantly acquiesce in British domination of the union should a true Europeanized politics fail to become preponderant, they certainly would not find French hegemony acceptable, and German even less so.

Moreover, the trend has been, and is likely to continue to be, for Germany to become stronger relative to France and the United Kingdom. It is possible that, as many Europeanists envisage, the latter two could cooperate to control the former, but their willingness and ability to do so are by no means assured. Although this possibility reduces the political uncertainty, it does not lower it to the point where the fears of Europeans would be stilled. At bottom, most Europeans are aware that the United States is their ultimate protector not only against the Soviet Union but also against a resurgence of German expansionism. Hence, they are not likely to sacrifice American support—whose conditions, however disturbing, they generally find tolerable—for the sake of membership in a new European superpower dominated by France or even by the United Kingdom, let alone by Germany.

In sum, the uncertainties regarding the issue of political control are major, if relatively unpublicized, factors in European decision making about the future of the unification movement. The lack of assurance regarding the political forces that would dominate an emerging European union is a basic consideration likely to continue to inhibit such steps in political unification as the establishment of a European nuclear force, a popularly elected European parliament with adequate legislative powers, and a more potent unitary, rather than conciliar, European executive. It would surely also deter agreement to confer on the EC's central institutions the necessary powers for controlling money and credit, for raising and spending substantial revenues, and for other economic functions that would constitute decisive advances in the movement toward full union.

The Effects on Unification of Relations with the United States

More ambivalent than any of the foregoing factors in their effects on the prospects for European unification are Western Europe's relationships with the United States. They are analyzed in detail in the next chapter and only two aspects need be briefly noted here.

The first major consideration is the relationship between European political-military unification and the U.S. nuclear guarantee. In the latter's absence, the European sense of the adequacy of the nation-state might again decline sufficiently to provide the necessary impetus toward unification in the military and political fields. But, there is a dilemma involved that hitherto has not been resolved in a way that fosters European unification. On the one hand, the United States cannot take the risk of explicitly removing its nuclear protection until the Europeans are clearly engaged in developing a credible nuclear deterrent of their own.[15] On the other hand, the Europeans need not divert substantial resources to this purpose and risk the political uncertainties of who would control the European nuclear force so long as the United States maintains its nuclear "umbrella" over them. Until this dilemma is resolved, continued reliance on the U.S. nuclear guarantee will contribute to inhibiting European political-military unification even though the size of U.S. forces in Europe is reduced and the latter's defense efforts are correspondingly increased.

The second consideration relates to the ambivalent effects on the unification movement of Western Europe's economic integration into the Atlantic region. Suffice it to say here that, in the course of the 1960s, the Atlantic region also achieved an unprecedented liberalization of trade and capital movements and an extraordinary flow of technology and managerial skills, which made major contributions to Western Europe's growth and prosperity. These benefits have involved, however, the reduction of freedom of action for national economic policies that is inevitable in any integrated arrangement. And, for the West European nations, such restrictions are felt to be all the more irksome because of the unique role played by the U.S. dollar in the international monetary system. The desire to escape the limitations thereby imposed on them has been an important motive impelling them to develop monetary arrangements of their own. The brusque and unilateral manner in which the United States instituted monetary and trade restrictions in August 1971, as well as the substance of those measures, greatly strengthened the European will to become more independent of U.S. policies. And, both the temporary resolution of these issues in December 1971 and the probable results of the subsequent negotiations for more basic agreements on Atlantic economic problems are likely to involve European institutional developments that will both express and sustain this conviction. Thus, in the years to come, the Community will be under continuing pressure, on the one hand, to preserve and increase the benefits it derives from Atlantic economic integration and, on the other, to reduce the restrictions on its freedom of action resulting therefrom by pushing ahead with its own unification.

In effect, the role of the United States in European unification is today, and will be in the foreseeable future, quite different from what it was in the 1940s and '50s. Whereas in the postwar period the movement toward European union would not have gotten underway without strong American leadership and support, so now even the reduced U.S. military presence in Europe and more qualified commitment to its defense constitute hindrances to the achievement of that goal. Nor do the increasingly important economic considerations operate predominantly to support the unification movement. Despite their strong urge for greater economic independence from the United States, most Europeans are, and are likely to continue to be, very reluctant to sacrifice the advantages they derive from a high degree of transatlantic economic integration.

The Will to Become a Superpower

The analysis so far has examined the main factors that wholly or in significant part counteract the self-reinforcing characteristic of European economic integration. Consideration must now be given to the possible forces that might sufficiently strengthen this tendency to assure eventual achievement of European union.

Since the mid-1960s, the Europeanists have been arguing that the Community's members were being increasingly confronted with certain imperatives that could be met only by transferring responsibility for them to supranational authorities. Some problems, like the technological and managerial gaps and the presumed threat of American private investment to European economic independence, express not only disparities in economic capabilities but also a large element of resentment and insecurity *vis-à-vis* the United States. However, these disparities have steadily been narrowing and they no longer constitute, if they ever did, a sharp spur to supranational development. In contrast, other problems, such as ecological dangers and related effects of rapid economic growth and urbanization, are increasing. But, even these difficulties have not reached a level of intensity that would make only a supranational approach to them effective. True, tackling them by means of intergovernmental cooperation and the coordination of national policies might be less efficient than by supranational authority. Unless in the coming years European governments perversely refrain from adopting the unilateral and joint measures required for dealing with them, however, these problems are not likely to become so urgent as to preclude use of the cooperative approach.

A much more potent force working toward European union would be a strengthening of the European will to become a superpower. The Europeanists argue that their nation-states are too small in terms of population and resources to achieve this status in world affairs, which could only be attained by a European union. True, a strong enough impulse to become a superpower would certainly revive the postwar sense of the inadequacy of European nation-states. The essential requirement for Europe to become a superpower is the strength of its will to do so, since the Community as a whole or any one of its three largest members already possesses the requisite technoeconomic capabilites. Whether or not it will have sufficient determination to pursue this course of development depends upon the answers to two questions.

The first is: what is the likelihood that European elites will become activistic and outward oriented to the required degree and that the people generally will go along with the necessary reallocation of resources from domestic welfare to military purposes? It seems probable that the attention and aspirations of European elites will become increasingly outward directed the more success they have in bringing about the internal institutional and other changes needed to achieve the domestic goals to which they have been according the highest priorities. Such developments might satisfy popular expectations sufficiently for the people generally to acquiesce in the reversal of the priorities between welfare objectives and playing a superpower role in world affairs. But, it is at least equally probable that the majority of the population would continue to oppose this shift because the proliferation of wants as resources increase is an inherent characteristic of affluent Western societies.

The second question is: assuming that most Europeans would acquiesce in such a reversal of priorities, will the elites and the people generally seek to play a major world role on an individual national basis, or through European union, or by arrangements for military and foreign-policy coordination that do not require a decisive transfer of sovereignty to supranational agencies? First, it must be reiterated that becoming a superpower is not the only way to play an important part in world affairs. Proto-superpowers can do so, too, as Chapter IV has explained. Second, as exemplified by General de Gaulle and his orthodox followers, the European elites most strongly committed to a great power role tend to be the most nationalistic, either opposed to European union or so intent on dominating it as to be likely to arouse the opposition of other members. Third, the fact that the cost-benefit ratio of becoming a superpower through European unification would be the most favorable would incline the majority of elites to this method of achieving such a status. Nonetheless, substantial savings could also be obtained under a looser, cooperative arrangement for coordinating military forces and foreign policies, although it would be less formidable as a world power. The individual national approach would be the most expensive in cost-benefit terms; hence, it would be the least rational way for the technocratic elites.

On balance, the cooperative approach would appear to have the highest probability. Although less efficient than uniting to become a superpower, it would still be sufficiently responsive to the European desire for greater independence of the United States and a major voice in world politics. This possibility is more strongly favored by the characteristics of the period of the new nationalism surveyed in the foregoing pages: the passive rather than active commitment to European union by the majority of elites and the people generally, the continued institutional strengthening of the nation-state, the political fears and rivalries impeding union, and the persisting ambivalence of the various considerations of rational interest involved in greater economic unification. Moreover, the longer these trends operate, the more powerfully will they inhibit the kinds of changes in attitudes and institutions needed to generate a sufficiently strong and widespread sense of the imperative necessity of attaining superpower status.

The Future of the European Community

There are three possible courses of development for the European Community over the foreseeable future: (1) to dissolve into its constituent nation-states; (2) to transfer enough of the crucial economic and political powers to the EC's central executive and legislative agencies to assure eventual completion of the unification process; and (3) to preserve and deepen economic integration and political-military coordination by means which do not involve a decisive increase of supranational authority in the Community.

As to the first possibility, the imaginable circumstances likely to cause the EC's dissolution have very low probabilities. One might be that the economic power, military strength and sense of mission of a major member—Germany, for example—would become so disproportionately great and the trend toward nationalized politics so accelerated and preponderant that domination of the Community by that nation would appear imminent. In an effort to avoid being trapped in the kind of relationship that enabled Prussia to control the mid-19th century Zollverein (customs union) and then to unite Germany politically under its rule, the other EC members might try to secede from it. Another unlikely possibility would be a voluntary agreement to dissolve the Community so that the members could obtain the greater benefits of participation in a much larger, emerging Atlantic union. The least likely way in which the Community might be terminated would be as the result of refusal to agree upon, or to meet the demands of a major member—France, Germany or the United Kingdom—regarding changes in the common agricultural policy, the steps toward monetary integration, or the adoption of common policies in other important fields. That negotiations over such matters will be difficult, prolonged and often bitter is highly probable in view of the experiences of the Community since the mid-1960s. However, it hardly seems within the limits of the possible that member governments would be so quixotic as to sacrifice the substantial advantages they are deriving from the customs union because they are unable to obtain the additional benefits believed to be at stake in negotiations for further integration. Even under de Gaulle, France did not behave in this fashion during the crisis of 1965; at worst, it boycotted many—although not all—meetings of EC bodies until a compromise was reached.

Regarding the second possible course of EC evolution, the analysis in this chapter leads to the conclusion that nothing is as yet evident in the development of the Community that has sufficiently strengthened the self-reinforcing tendency of economic integration to give an economic and political union the highest probability. In the existing circumstances, the critical test is likely to come in the second half of the 1970s over the institutional changes required for continued progress toward monetary, and hence full economic, union. If agreement is reached to give the EC's central institutions the power to regulate money and credit and the ability to harmonize national economic conditions by ordering modifications in national budgetary and fiscal policies, then achievement of a true European union would be the most probable course of development. For, to carry out those responsibilites effectively, the supranational authorities would sooner or later also have to be granted the additional power to levy significantly large taxes and to dispense the resulting revenues without national-government concurrence. And, in that case, the probabilities would be enhanced that the supranational agencies would eventually obtain the power to control the armed forces and to conduct the external relations of the, by then, clearly emerging European union.

As matters now stand, however, the possibility with the highest probability is that these decisive steps will not be taken but that the Community's economic integration and political-military coordination will be broadened and intensified in a variety of other ways.

With respect to economic integration, the most conspicuous developments in the shorter term will probably be in the monetary field. However, they are not likely to involve decisive progress toward the common currency envisaged in the plan of monetary union adopted in 1971. Although members have agreed to narrow the fluctuations among their currencies, it is improbable that exchange rates among them would be permanently fixed and interconvertibility made unqualified and irrevocable. The reason is that the European central bank (or its equivalent) or the mandatory coordination of national economic conditions and policies required to make such a system work would necessitate more supranational authority and greater impairment of national freedom of action than are likely to be acceptable to the majority of elites and the people generally, especially in France, Germany and the United Kingdom. Also, rivalry among sterling, the mark and the franc, as well as reluctance to assume the responsibilities involved, will tend to inhibit use of any one of these qualified member currencies for intervention and reserve purposes under a system of permanently fixed rates and freely interconvertible currencies. Rather, the probability is greater that, in the next few years, more reliance will be placed upon voluntary cooperation to harmonize national economic conditions and, particularly, to coordinate European monetary policies *vis-à-vis* the United States and in the negotiation of international monetary reforms.

In addition to monetary measures, other steps toward further economic integration that have a good chance of being implemented would include revision of the common agricultural policy to accelerate the rationalization of European farming and reduce its financial burden; common policies for such areas as energy, transportation, communications, patents, company law; harmonization of business taxation, social-welfare charges and benefits, product and safety standards, government subsidies and procurement practices, regional development programs, and other so-called "nontariff" barriers and distortions significantly affecting the conditions of competition. The more integrated the Community becomes in these and other respects, the freer market forces will be to harmonize economic conditions among the member states, thereby easing the difficulty of keeping exchange-rate fluctuations within narrower margins through voluntary coordination of national policies and the provision of short- and medium-term credits. Nor would such closer monetary cooperation preclude—as would a common currency—occasional changes in member-countries' par values as an alternative to unacceptably drastic internal adjustments.

These economic developments would foster, and would be reinforced by, the parallel process of political-military coordination. Joint technological and military research programs and more integrated arrangements for defense production would lead to greater coordination of armed forces. These changes would be likely to begin with growing cooperation between the British and French nuclear forces, discussed in Chapter IV, which would probably be the basis for developing a credible second-strike capability with the technoeconomic assistance of the other EC countries, especially Germany. Depending on circumstances, Germany might eventually be permitted to share with France and the United Kingdom in the top-level decision making regarding this European nuclear force. Closer military integration would, in turn, necessitate greater coordination of foreign policies, which would be a precondition for a more active and independent European role in world politics.

Although they would not represent decisive progress toward economic and political union, such advances in political-military coordination combined with the intensification of economic integration would increasingly constrain the EC's member states to think of themselves and to act as a unit. These developments would both ease Europeans' ambivalent attitudes toward the United States and enhance the Community's ability to deal with the United States on a more nearly equal basis. Finally, the Community's association agreements with other European countries, as well as with Mediterranean and African nations, would tend to give it a growing importance in world economic and political affairs.

In this projection, these economic, political-military and psychological changes would sooner or later be expressed in modifications in the Community's central institutions. While not likely to involve the transfer of the crucial economic and political-military powers to the supranational authorities of a true federal union, they would result in the formal establishment of a European confederation. This arrangement might be reached by the course proposed by President Pompidou, or it could evolve in other ways and take a different form.[16] Essentially, however, the critical sovereign powers would be retained by national governments even though they would allocate substantial revenues to the confederal institutions and delegate to them important representational responsibilities in external economic and political negotiations.

This course of development would fall significantly short of giving the Community the cohesiveness needed to make it a worldwide superpower comparable in military strength and sense of mission to the United States and the Soviet Union. Nevertheless, its economic importance as the world's largest trading unit and its ability to act in a concerted fashion to advance or protect its interests—if not to express a mission to transform the international system—would make it second only to the superpowers in the capacity to influence the behavior of other nations. In a sense, the confederal Community would become the premier proto-superpower, approaching but never quite attaining superpower status.

As in the past, there will continue to be considerable confusion among U.S. policy makers and opinion leaders over the significance of EC developments owing to the widespread failure to distinguish between the different, though related, processes of economic integration and economic and political unification. Although official spokesmen of the Community may sometimes be overly optimistic in interpreting the longer-term implications of measures proposed or adopted, they have generally not been deliberately misleading. Nevertheless, such terms as "European union," "federal union," "united Europe" will continue to be prominent in the explanations and justifications of European publicists and politicians. Governmental and private policy makers and opinion leaders in the United States need to be careful not to take such statements at face value but to determine whether the real significance of EC changes lies in the deepening of economic integration and political-military coordination or in the advancement of economic and political unification. It is important for them to do so because whether the Community follows a federal or a confederal evolution would lead to different kinds and degrees of tensions between it and the United States. Conversely, the problems of transatlantic relations, which in part reflect the perceptions and conceptions of American policy makers and opinion leaders, will be major influences on the course of European development, as explained in the next chapter.

Transatlantic Relationships in the Period of the New Nationalism

How is the Atlantic regional system likely to evolve in the foreseeable future? Will it develop into an arrangement sufficiently resembling an Atlantic partnership for this long-time goal of American policy to be in essence achieved? Or are there other, more probable outcomes of existing and prospective trends that would lead to quite different kinds of relationships? This chapter endeavors to answer these questions by analyzing the highly complex and often contradictory factors involved. But, before turning to these subjects, two general points need to be emphasized.

Basic to the development of the Atlantic regional system is the fact that its constituent nation-states are descendants of Western civilization and comprise those contemporary forms of Western society and culture that have the greatest similarities in institutional systems, values and behavioral norms. Of all the ties that bind the Atlantic nations together, those comprised in this common sociocultural heritage are the most pervasive and powerful. And, in the period of the new nationalism, when for the first time world politics has become truly global and all countries participate in it as more or less significant actors and no longer as mere spectators or helpless prizes, the importance of socio-cultural similarities and differences will be manifested in many unanticipated ways.

The sociocultural affinities are crucially important in another respect, too. The interplay of trends and countertrends, the inconsistencies in values and interests, the contradictions in attitudes and relationships characteristic of Western society are also expressed in the conflicting integrative and divisive forces shaping the Atlantic regional system. The prospects of the region cannot be adequately understood except in the context of the deep ambivalences that exist not only within Western Europe, as explained in the preceding chapter, but also in the Atlantic region as a whole. The two levels of development are in part inversely related: the more united economically and politically Western Europe becomes, the less integrated economically and politically the Atlantic region will be, and *vice versa.* But, although at the extremes they are mutually preclusive, throughout the large intermediate range they are, paradoxically, both dependent on and incompatible with one another. The more probable of the courses of European and Atlantic development lie within this intermediate range. The paradox results in part from the disparities in the relationship between Western Europe and the United States and in part from the inconsistencies between the pressures of the new nationalism and the constraints of regional economic integration.

Transatlantic Relationships and Attitudes

During the postwar decades, the most striking characteristic of the transatlantic relationship was the extreme dependence of one side on the other. Both economically and militarily, Western Europe's survival and future prospects were substantially affected by the policies and actions of the United States. This dual dependence was symbolized by the Marshall Plan and NATO, whose origins and significance were sketched in the preceding chapter. Indeed, as explained there, Western Europe's reliance on the United States went beyond its need for economic resources and military protection. The most important European development of the postwar period, the creation of the European Community in the late 1950s, was itself dependent upon American leadership and support.

Throughout the postwar period, the West European nations related to the United States predominantly on a bilateral basis. In consequence, each saw itself confronted with an American economy many times larger than its own and characterized by competitive practices and innovative tendencies of an unfamiliar and frightening intensity. These gross economic disparities were paralleled by the wholly disproportionate political and military capabilities of the United States, then the premier superpower. Moreover, Europe's weaknesses were so marked and the dangers of the cold war appeared so menacing that the disparities in the transatlantic relationship tended to be accepted as irremediable facts of life. European gratitude for the liberation from Nazi conquest and for the unprecedented size and liberal terms (four-fifths were grants) of Marshall Plan aid helped also to sustain U.S. prestige and acceptability.

In those years, U.S. economic and political power provided means for active expression of the American sense of mission, which sought to hasten the unification of Western Europe so that it could become one of the "twin pillars" of an Atlantic partnership—in turn, regarded as a precondition for transforming the international system. For their part, most European policy makers and opinion leaders shared in some degree the U.S. conviction of its world-redemptive mission and the American conception of a more rational world order. In consequence, European faith in U.S. goodwill and European confidence in American leadership were strong enough to repress the resentments and frustrations inevitably engendered by so disparate a relationship.

With the completion of European economic recovery and the national, regional and international changes associated with the transition to the period of the new nationalism in the course of the 1960s, the transatlantic relationship became more overtly and disturbingly ambivalent. By the beginning of the decade, the West European countries were economically strong and prosperous enough to participate in the process of Atlantic economic integration, analyzed in the next section. Europeans were then unhappy to discover that their newly acquired capacity for independent policies was increasingly restricted by the effects of the disproportionate size and freedom of action of the American economy. Unlike the situation in the postwar period, when their needs were met by U.S. aid requiring little, if any, sacrifice of American freedom in domestic and foreign affairs, Europeans now wanted substantial changes in the U.S. national policies affecting their well-being—concessions that Americans were neither willing nor able to grant. Desirous of obtaining the benefits of American investment and technological and managerial innovations, Europeans were alarmed at the more aggressive approach to competition of American business firms and at the extent of their interest in establishing or acquiring subsidiaries in Europe. The resulting feelings of frustration and resentment were expressed during the 1960s in recurrent complaints about the effects of U.S. balance-of-payments deficits, fear of domination by large American corporations, and alarm over the so-called "technological, managerial and educational gaps" and the "brain drain."

These resentments and fears generated by economic disparities were reinforced during the transitional phase to the new nationalism by the dissatisfactions and suspicions arising from political and military disparities. On the one hand, because their own interests were at stake—

including, most basic of all, that in world peace and war—Europeans wanted an effective voice in, or at least a veto on, U.S. policy making for NATO and for dealing with situations elsewhere in the world that might involve them willy-nilly. On the other hand, they were unwilling to allocate resources for developing the conventional and nuclear military capabilities that alone would have given them the power to insist upon having such a voice. Nor was the United States inclined to share major policy making with them for fear of impairing its own freedom and speed of action in critical situations. The American proposal for enhancing European status in the alliance through the establishment of a multilateral nuclear force (MLF) generated little support because it would have been effectively under U.S. control, was unresponsive to the European political limitations discussed in Chapter V, and would have necessitated a much greater commitment of European resources. Equally revealing of European dissatisfaction was the fact that no European country was willing to allocate money or men to assist the U.S. effort in Indochina. Indeed, most European governments refused to give even a verbal endorsement to U.S. intervention and some openly opposed it. Nothing contributed more during the 1960s to the decline of American prestige and of European faith in U.S. leadership than the Indochina involvement.

With the onset of the period of the new nationalism at the beginning of the 1970s, transatlantic relations entered their third and current phase. Completion in mid-1968 of its customs union made the Community the world's largest trading entity, responsible in 1971 for more than 17 percent of total world exports (excluding trade among EC members) compared to less than 15 percent for the United States. If the United Kingdom, Denmark, Ireland and Norway had been members in 1971, the enlarged Community's share of world exports would have risen to around 23 percent (excluding trade among EC members and the four additional countries) while the U.S. share would have increased by less than one percentage point.

Moreover, the importance of the Community in the international economic system was further enhanced by the proliferation of association agreements of various kinds with nonmember countries. Originally, such arrangements for abolishing or substantially reducing tariffs and other trade barriers on a mutually preferential basis were instituted by the Community with the newly independent African countries formerly colonies of its members. The purpose was to assure that their raw-material exports would continue to have a privileged status in the Community and, conversely, that EC exports would retain a similar advantage in their markets. Subsequently, agreements of this type were signed with the African countries formerly under British rule. Also, in the course of the 1960s, the Community acceded to pressure for similar preferential treatment from one after another of the South European, North African and Levantine countries, all of which were substantially dependent on agricultural exports to the Community and each of which feared the trade-diverting effects of the EC's agreements with its competitors. Finally, the EFTA countries that did not join the Community sought association agreements with it in order to preserve freedom of trade in industrial products with their former EFTA partners and to gain similar free access to the vast and rapidly growing market of the rest of the Community. Thus, by the early 1970s, all of Western Europe, most of the other Mediterranean nations, and all of black Africa (except Ethiopia and Liberia) were bound more or less tightly into a preferential trading and investing area centered on the European Community.

These developments augmenting the economic power and influence of the Community were reinforced by the psychological effects of continuing economic expansion and rising living standards. In turn, they led to increased European self-confidence and a growing sense of having a collective, as well as an individual, relationship with the United States. This change significantly assuaged the anxieties generated by transatlantic economic disparities and lessened, although it did not eliminate, the ambivalences in European attitudes toward the United States. Many prominent manifestations of European resentment during the 1960s—such as the agitation over technological and other "gaps" and the fear of domination by American multinational corporations —gradually diminished. At the same time, however, other conflicts inherent in transatlantic economic, political and military relationships in the period of the new nationalism became increasingly important. Before analyzing these problems, existing and prospective attitudes toward the transatlantic relationship in Western Europe and the United States need to be taken into account.

Although the attitudes of people generally in Western Europe toward the United States are today less ambivalent than those of the elite groups, they are nevertheless still quite complex. The majority of Europeans are aware that their security and freedom rest ultimately on the nuclear deterrent of the United States. True, they tend to believe that the cold war is over; but they still do not trust the intentions of the Russians or of the Germans. So long as there seems to them to be a reasonable chance that either of these nations may once again try to dominate Europe, and granted their continuing unwillingness to allocate resources to national or European defense at the expense of their economic and cultural goals, the majority of West Europeans want to maintain a close enough relationship with the United States to assure American help in time of need.

Moreover, these considerations of rational interest are reinforced by a vague but nonetheless deeply rooted faith in America on the part of large numbers of Europeans. In origin, it was the counterpart in the many people who remained in Europe of the feelings and expectations, described in Chapter III, that induced those Europeans who sought a new beginning to emigrate to the United States. Just as for the latter, America was the land of promise where aspirations for freedom and plenty could be realized, it also became so for the former, especially when, after the turn of the century, the preeminent power and wealth of the United States were increasingly apparent. These general feelings in Europe were strengthened by the flow of emigrants' remittances, by the comparative affluence of returning migrants, by the idealism of Woodrow Wilson and the humanitarianism of Franklin D. Roosevelt, and by the crucial American assistance in two world wars and the large-scale aid provided by the United States for wartime relief and postwar recovery. In turn, these experiences validated in European minds America's own image of itself as a moral, sociopolitical and technoeconomic paradigm, and they have been responsible for the European tendency to judge American actions by more rigorous ethical standards than those applied to the Soviet Union—or to themselves.[1]

Hence, despite the envy aroused by disproportionate American wealth and power, the worry provoked by reports of America's domestic crises, and the fear that American redemptive activism might incite a third world war, desire for an American commitment to help Europe preserve its freedom and prosperity continues to be widespread. Yet, while they believe it in their interest to maintain the tie with the United States, the great majority of Europeans now want to be treated as fully independent, self-responsible people, not as mere clients in an American hegemony. This attitude has two consequences. On the one hand, it means that Europeans react strongly against American efforts to press them to adopt domestic or foreign policies that they do not favor of their own accord. On the other hand, the splitting of the Atlantic region as a result of growing dissatisfactions and resentments would be likely to arouse deep feelings of anxiety in most Europeans, and such a development would be met with disquieted resignation rather than enthusiastic support.

The attitudes of European elite groups toward the United States tend to be markedly more ambivalent than those of the people generally. Such conflicted feelings are resolved with least difficulty by the technocrats. After all, American society and culture contain the pioneering and quintessential exemplars of the rationalized contemporary institutions and positivistic ways of seeing, believing, thinking and acting that also characterize those of Europe. This is often interpreted as the "Americanization" of Europe and condemned as evidence of American economic or cultural "imperialism." These surface likenesses lead to the erroneous conclusion that the chronologically earlier development must be the cause of the later. In fact, as explained in Chapter II, the roots of both lie in the same underlying processes of socioeconomic change and cultural evolution occurring in all Atlantic societies, and in the common conceptual and institutional needs of all highly complex, internally and externally interdependent, industrialized market systems. Hence, European and American technocrats in government, business, and universities and other organizations tend to "talk the same language," worry about the same kinds of problems, use the same rational and technical means for dealing with them, and motivate their actions by the same positivistic expectations that reason and science will control nature and guarantee the progress and eventual perfection of society. In turn, these occupational and intellectual affinities are responsible for the empathy, cooperation and mutual emulation and competition that exist among the technocratic groups throughout the Atlantic region regardless of national boundaries.

Moreover, the majority of European technocrats are well aware of the rational interest of their countries in maintaining good relations with the United States. Unwilling in the shorter term to devote substantial resources to building up their own nuclear and conventional military forces, they are resigned—however reluctantly and uncertainly—to continued reliance upon the American nuclear deterrent. They appreciate the value of the high degree of economic integration in the Atlantic region and the importance for their countries' economic growth of the transnational integration of production, described in the next section, with its increasing two-way transatlantic flow of goods, capital, managerial talent, and scientific and technological innovations. At the same time, these common interests do not prevent conflicts over particular economic and political issues. Nor do they preclude severe and often contradictory criticism of American foreign policy: on the one hand, for its excessive universal scope, missionary zeal and activism believed to endanger world peace; on the other, for its complacency about Russian and other threats to European freedom and welfare and its pursuit of a bilateral *détente* with the Soviet Union, presumably at the expense of European interests.

Most European technocrats are by now sufficiently confident of their own knowledge and skill and of American self-restraint to have little, if any, fear of outright political or economic domination by the United States. But, they are wary of the effects of disproportionate American wealth and power on their own freedom of action, and they feel a corresponding need for vigilance in detecting, and firmness in resisting, U.S. efforts to influence their opinions and policies. Even before the progress of internal modernization and improvement makes it possible for them to allocate resources to achieving proto-superpower status, European technocrats—like the people generally—want the United States to treat their nations as independent, self-responsible participants in world affairs.

These attitudes of the majority of European technocrats shade gradually into the deeper ambivalences of the younger and more ardent Europeanists. Neither on the continent nor in the United Kingdom are most younger Europeanists, who aim at a superpower role for a united Europe, motivated explicitly by anti-American feelings—although a minority certainly is.[2] However, they recognize—as many American politicians, officials and publicists still do not—that the interests and attitudes of such an activist united Europe are unlikely to make it support automatically the objectives, policies and operations that the United States believes are in the common good.

American attitudes toward Europe have also been in process of change but they are becoming more, not less, ambivalent. On the one hand, Americans continue to feel greater common affinities and goodwill toward Western Europe than toward any other region. On the other hand, the developments of the 1960s have led to the emergence of genuine conflicts of interests and objectives across the Atlantic, whose negative effects on American opinion are aggravated by the disappointment of unrealistic expectations of how the Europeans would behave regarding these issues.

Although the rhetoric of the Grand Design of Atlantic partnership is no longer officially used, many U.S. policy makers and opinion leaders continue to assume that both rational considerations and moral obligations will preclude serious transatlantic divergences. Europeans are expected to conceive their interests in the same terms as Americans and hence to cooperate willingly in the common good. The growing incongruities between unfounded expectations and unwelcome realities tend to be blamed not on American misconceptions but on the ir-rational short-sightedness and perverse ingratitude of Europeans. In part, these American reactions are rationalizations of self-serving mea-sures that interest groups and opinion leaders in the United States are seeking, such as protective tariffs and quotas. But, in part, they are genuine reflections of the ambivalences of American attitudes toward Europe arising from redemptive activism and sentimental or ration-alistic parochialisms. As such, they both exaggerate the seriousness of legitimate American concerns about self-interested European behavior and impair the ability of the United States to deal effectively with these problems.

The effects of these conflicting feelings on American attitudes toward Europe have been reinforced by other changes of the late 1960s and early '70s affecting the relative position of the United States in the international system and the shifting goals and priorities of American policy. Just as diminishing European ambivalences *vis-à-vis* the United States express growing European economic capabilities and self-confidence, so the relative decline of American political and military power and the self-doubts generated by the frustrations of the Indo-china involvement and the seeming intractability of domestic difficul-ties make the United States more self-concerned and less forbearing toward its allies and trading partners. Americans increasingly resent the unwillingness of prosperous Europeans to meet the burdens of European defense, and they are becoming more insistent on protecting or advancing their own interests relative to those of the Community and Japan. U.S. policy makers and opinion leaders are more deeply concerned over the competitive position of the American economy and, therefore, more resentful of the EC's preferential trade arrangements and the restrictive practices of the Japanese. Thus, while Europeans are becoming less fearful of and deferential toward the United States, Americans are becoming more wary of European—and Japanese—capabilities and more critical of their real and imagined deficiencies.

The conflicting perceptions and conceptions that permeate both sides of the transatlantic relationship are important less because they occasionally manifest themselves in emotional outbursts than because they continually color and distort economic and political issues and intensify the sense of their urgency and magnitude. Moreover, the ambivalences of attitudes interact with the other equally complex set of inconsistencies that also plays a major role in the development of the Atlantic region—those between regional economic integration and national needs and limitations.

The Nature and Significance of Atlantic
Economic Integration

The most conspicuous feature of the Atlantic economic system is the high degree of integration that emerged in the course of the 1960s. In part, economic integration reflects the absolute growth of trade and capital movements within the Atlantic region despite the persistence of various national barriers (such as tariffs and quantitative restrictions, foreign-exchange and investment controls, discriminatory taxation and government procurement practices and subsidies). In part, it results from the very substantial reduction of many of these governmentally imposed restrictions on trade and payments.

The interest of the Atlantic nations in maintaining and extending their economic integration arises essentially from the fact that it is a major means for achieving economic growth.[3] Economists identify both "static" and "dynamic" effects of integration on growth.

The first arises from the well-known economic principle of the mutual gains from foreign trade and investment in accordance with the comparative advantages of the national economies involved. Many Atlantic countries depend upon foreign commerce to help maintain their living standards both because of their need to import raw materials lacking in their natural resource endowments and other goods they are unable to make at economical costs, and because exports constitute a substantial percentage of their gross national products (GNP). The benefits to employment and incomes of importing and exporting capital have also been substantial.

The contemporary changes in institutions, values and attitudes make the dynamic effects more important. Difficult to measure directly, they embrace the various ways in which the freer and bigger flows of goods and capital in an integrated regional system stimulate and sustain the growth rates of its constituent national economies. The enlarged market, made available by the openness of comparatively small national economies to one another, provides opportunities for new investment and for improving productivity through both the internal economies of scale and the external economies of easier access to cheaper or more diversified ancillary goods and services of all kinds. Equally important are the more intangible and pervasive effects subsumed in regional competition. In addition to the stimulus of competitive imports of goods and services, they include the dynamic effects of competitive development of new products and production and marketing techniques, of competition in devising and applying new organizational arrangements and management methods, and of rivalry to be the first to enter a new market or branch of industry and to be the biggest or the leader in a particular field of production, distribution or finance. In these and other ways, regional competition fosters the self-confidence, initiative, innovation, entrepreneurial vigor, flow of ideas and technologies, and flexibility that are among the major psychosocial components of economic growth in pluralistic societies.

Even for the nearly self-sufficient American economy, whose 50 states still constitute the biggest freely trading market on the planet, the opportunities and competitive pressures resulting from its integration into the Atlantic region were important impulses to maintaining its dynamism during the 1960s. And, in turn, the various stimuli and competitive influences radiating throughout the region from the United States have been among the most significant factors contributing to the high growth rates of other Atlantic nations—as well as a major source of problems and complaints.

The Progress of Atlantic Economic Integration

The measures by which, since World War II, Atlantic economic integration has gradually been achieved can be briefly summarized. Even before the war ended, the Bretton Woods Conference of 1944 designed the International Monetary Fund (IMF) to assist in restoring and maintaining an effectively functioning international monetary system. A conference at Havana in the winter of 1947/48 envisaged that the IMF would be paralleled by an international trade organization, but the resistance of the U.S. Congress was largely responsible for the failure to establish it. Instead, the General Agreement on Tariffs and Trade (GATT) was drawn up to specify the rules and procedures for a worldwide, nondiscriminatory, multilateral trading system. Later, the GATT was developed into an institutional means for reaching and jointly scrutinizing the implementation of agreements to reduce or abolish the neomercantilist practices—such as protective tariffs and quantitative restrictions, bilateral agreements and preferential trade arrangements, export subsidies and dumping—adopted during the great depression of the 1930s and the wartime and immediate postwar years of the 1940s. The third continuing intergovernmental institution is the Organization for Economic Cooperation and Development (OECD) established in 1961 as the successor to the OEEC. In addition to the latter's European members, the OECD included the United States and Canada from the beginning and has subsequently been joined by Japan (1964), Finland (1969) and Australia (1971). It provides an increasingly important forum for periodic review of national measures affecting economic growth and regional integration, and explores the kinds of policies and programs required by its members to keep up with educational, scientific and technological developments.

In consequence of the economic revival stimulated by the Marshall Plan, the general realignment of exchange rates in 1949 and subsequent individual devaluations, the good rates of economic growth maintained during the 1950s, and the increasing international availability of dollars resulting from persisting U.S. payments deficits, the West European nations were able to restore the current-account convertibility of their currencies at the end of 1958. Equally significant was the fact that six rounds of tariff-cutting negotiations under GATT auspices, culminating in the Kennedy Round of 1962–67, resulted in a drastic lowering of tariffs affecting trade in industrial products among all of the Atlantic countries, as well as with GATT members in other parts of the world. Since the tariff cuts agreed upon under the Kennedy Round became fully effective on January 1, 1972, the Atlantic region has had a lower level of tariff restrictions on nonagricultural products than existed before 1914.

Trade liberalization was paralleled by the gradual freeing of short-term capital movements within most of the Atlantic area. In addition, many European countries, notably the United Kingdom and Germany, liberalized their controls on long-term capital movements. However, since 1958, the most significant portion of growing long-term capital flows within the Atlantic region has been the direct investment of American private capital in European industrial, financial and other activities. This has been matched by the movements of long-term European capital to the United States, mainly into portfolio securities but, in recent years, increasingly into direct investment as well. As of the end of 1970, the total accumulated transatlantic long-term private investment was roughly in balance, with about $30 billion of American holdings in Europe predominantly direct and around $32 billion of European holdings in the United States still mainly portfolio.

The trend toward direct regional investment not only by American companies but increasingly also by European firms reflected several developments and motivations. The first were, of course, the opportunities arising from the high growth rates and increasing purchasing power of Atlantic economies and from the enlarged market areas provided during the 1960s by the two free-trade arrangements in Europe, especially the economically bigger and geographically more concentrated European Community. Another was the greater attractiveness for American corporations of manufacturing within the Community and the EFTA compared with trying to export from the United States over their remaining tariff and other barriers against nonmembers. This advantage was reinforced by the savings on transportation and other costs of producing closer to markets, and by the difficulty of most American companies, oriented toward the gigantic U.S. home market, of devoting adequate attention and personnel to relatively much smaller export operations. Other important considerations were the desire of American and European companies to keep capital accumulations more profitably employed abroad than they could be at home; and the pressure on American business firms during the 1960s to "follow the leader" to Europe not only for economic but also for prestige reasons.

In part created by and in turn helping to make possible the high levels of direct regional investment is another major manifestation of Atlantic economic integration: the rapid growth of the Eurocurrency market during the 1960s. It is comprised mostly of dollars, augmented by smaller percentages of readily transferable European currencies, deposited at interest in European branches of American banks and in European banks. The main sources of Eurodollars are portions of the official dollar reserves of Atlantic governments, the funds of American and other companies required for or resulting from their expanding direct investments in Europe, the proceeds of exports by European and Japanese firms to the United States, and capital from outside the Atlantic region, especially from the oil-rich Moslem countries, Latin America, and the overseas Chinese communities in East Asia. The total amount of credit, net of redeposits, extended through the Eurodollar market from these sources was estimated at around $50 billion at the beginning of the 1970s. Eurodollars and the other Eurocurrencies are now the equivalent of a freely moving international capital market for the Atlantic region analogous to that of the 19th-century system.

The Transnational Integration of Production and the Spread of Multinational Companies

These characteristics of the contemporary Atlantic system would seem to indicate that it differs largely in degree from the integrated world economy of the second half of the 19th century. However, implicit in them are certain crucial qualitative differences in structure, institutions and policies that are already of major significance and are likely to be increasingly important in the future.

The structural changes comprise the various aspects and implications of what may be called the distinctive trend toward *transnational integration of production* in the Atlantic region. In the 19th-century system, trade among nations consisted mainly of raw materials (ores and refined metals, fuels, cereals and other agricultural products) and finished commodities for consumption or, as in the case of most textiles, for direct conversion into consumers' goods by handicraftsmen and households. In the contemporary Atlantic system, of course, trade in finished consumer products has increased enormously owing both to the high rates of economic growth in the countries concerned and to the lowering of barriers among them. However, compared to the 19th-century system, two other classes of goods are now much more important. They are capital goods (production machinery and ancillary equipment of all kinds) and intermediate commodities (semiprocessed metals, manufactured chemicals and synthetic materials, and an immense and growing variety of parts, subassemblies and other types of components needed to make all kinds of final capital and consumer products). The substantial proportion of total regional trade now in these two categories is a manifestation of the growing interdependencies of the various stages of the production process across national boundaries. This trend exists not only within the European Community and between contiguous Canada and the United States but also between North America and Western Europe, and between all of the latter and Japan and Australia. These developments have been facilitated by improvements in the speed and efficiency of ocean transportation.

This change in the composition of regional trade reflects, and in turn helps to foster, the parallel changes in the composition of regional investment and in its relationship to trade. In the pre-1914 system, the regional flow of long-term capital was almost completely into portfolio investments in the securities of other Atlantic countries, and regional trade was several times larger than the output of direct foreign investments, that is, of manufacturing, financial and service enterprises owned and operated by foreigners. Although long-term European holdings in the United States are still predominantly portfolio investments, the subsidiaries in the United States of European, Canadian and Japanese companies have been growing in number and size. For their part, long-term American holdings in Canada and Europe are overwhelmingly in direct investments. In reversal of the 19th-century relationship, the total annual output of American, European, Japanese and Canadian subsidiaries in other Atlantic countries substantially exceeds total regional trade. Because American direct investment is so large, total production of goods and services by the subsidiaries of American companies in other Atlantic countries is several times bigger than U.S. exports to them; indeed, it constitutes a substantial portion of Canada's GNP and significant percentages of the GNPs of the leading West European nations.[4] And, a large share of the growing regional trade in capital goods and intermediate products is conducted *within* American, European, Japanese and Canadian corporate entities or groups—that is, among parent companies and their various subsidiaries or affiliates in other Atlantic countries.

Because direct investment by American companies throughout the region (except in Japan) was so prominent a feature of the 1960s, it still tends to be regarded as the only significant manifestation of the transnational integration of production. In the European Community, the flow of the various kinds of products within and among European firms is also very important. Indeed, the growing network of ties among independent European companies in different EC countries—through long-term suppliers' contracts, marketing and servicing agreements, licensing of patents and processes, ownership of each other's securities, mutual credit extensions, and so forth—has hitherto constituted a substitute, as well as a reason, for the conspicuous lack of formal mergers among such firms. In addition, of course, many of these companies have been establishing subsidiaries in other EC countries, but this direct investment still is less important than the network of relationships among independent European corporations.

The predominant institutional form in which the transnational integration of production is expressed is the *multinational enterprise*. This is the latest development in the evolution of the modern corporation sketched in Chapter II. Like previous innovations in corporate organization and methods, it has major social implications—especially for the international system—the most important of which may still lie in the future.

Both names for and definitions of the multinational enterprise vary, but the term generally refers to large corporations with branches or subsidiaries in an increasing number of countries in addition to those in the home nation of the parent company. By this definition, multinationality is not a new characteristic; as early as the last quarter of the 19th century, there were already European and American companies with substantial foreign investments not only in extractive activities but also in manufacturing industries. However, by the mid-1960s, the quantitative growth in the number, size, diversity and geographical extent of multinational enterprises was so great as to constitute a qualitative difference in the nature and significance of this development. The difference arises essentially from the fact that, in multinational enterprises, an expanding share of assets is located in foreign countries; a rising percentage of production and sales takes place outside their home markets; growing proportions of the labor force and managerial and technical personnel are of different nationalities; and more and more of the loan portion of their capital, especially for their overseas investments, is raised in foreign money markets. As yet, however, such a trend is not nearly so marked in equity ownership and is barely discernible in the promotion of managers of foreign origin to top policy-making positions in the parent company, but these additional indicators of multinationality are likely to begin to increase more rapidly in the course of the 1970s.

The consequences of these changing proportions are that the conceptual framework of multinational corporations is expanding beyond national—indeed, in most cases, beyond regional—horizons; the strategic options open to their decision makers are increasing; and their vulnerability to adverse developments in any particular market or country is being reduced. These broadening perspectives and possibilities more and more affect corporate decision making regarding product and production planning, the location of new investments and research facilities, transfer pricing, financial management of liquid funds, health and environmental protection standards, personnel practices, customer and community relations, and other aspects of policy and operations. In making these decisions, certain factors of determinative importance in the past—such as tariffs, quota restrictions, exchange and capital controls, are less significant today. Others have become correspondingly more important—for example, labor supply and unit labor costs, nearness to and cost of other required production inputs, closeness to major customers and areas of high sales concentration, transportation costs, and various other locational advantages and external economies. Moreover, governmental policies—taxation, regulation of business activities, procurement practices, subsidies and other aids, local development programs, and others—as well as political conditions and attitudes toward private enterprise in general and multinational companies in particular, continue to be major considerations. Thus, the transnational integration of production does not mean that national differences have ceased to be important elements in operating decisions and long-range planning in multinational enterprises, as some commentators have concluded. Quite the contrary: the reduction or disappearance of tariffs and other trade barriers *per se* makes differences in national economic conditions and attitudes and in other government policies relatively more significant.

The transnational integration of production manifests itself not only in these structural and institutional changes but also as a cultural phenomenon. This consists of the already vast and still increasing number and variety of economic communications of all kinds, the greater and faster flows of familiar and novel information across national boundaries occurring within and among business firms and other private organizations. Of particular importance is the rapid diffusion of new technologies, managerial methods, and other types of knowledge that have major effects on productivity and competition. The increased flow of information and ideas both results from and reinforces regional economic integration, fostering the harmonization of national economic conditions and policies, and reflecting and further intensifying the transnational integration of production.

In addition, the transnational integration of production and the rise of multinational enterprises have important monetary effects. The growth of trade in all categories of goods means that payments within and between companies across national boundaries are greatly increased both absolutely and relative to the size of the business, as measured by total sales or in other ways. The volume and velocity of regional money flows are further magnified by the large amounts of funds kept abroad. The transnational scale of their activities and interrelationships requires North American, European and Japanese companies to maintain bigger working balances in other Atlantic countries, and generates growing accumulations of profits and depreciation funds that sooner or later are either remitted to parent companies or reinvested abroad. Capital transferred from parent companies or raised in the Eurodollar and other mobile currency markets to start new subsidiaries or to purchase or expand existing enterprises may be temporarily unneeded for these purposes, and hence may also be available for short-term use. Finally, banks and other types of financial institutions have established more branches and affiliates throughout the region, and shift balances freely and in much greater volume than in the past among them and between them and their home offices. All of these funds tend to move rapidly throughout the region to obtain the most advantageous returns. The effects on the international monetary system of this significant increase in the size and volatility of money flows among Atlantic countries are discussed in a later section of this chapter.

Finally, transnational integration has intensified competition within the Atlantic region. The rising scale and importance of economic activities and communications across national boundaries inevitably means increased competition among the business firms, as well as the farmers and wage-earners, of the Atlantic countries. The effects at both *micro* and *macro* levels are complex, reflecting not only economic but also political and psychological factors, and tend therefore to be highly controversial, as explained in a later section.

The Ambivalences of Regional Integration

The contemporary Atlantic economic system differs from the integrated 19th-century economy not only in the structure of international trade and investment and in the nature of its most significant institutional form but also in the means by which integration is maintained. Owing to differences in attitudes and expectations, however, the effects are deeply ambivalent in the existing system whereas they were not in the 19th-century world economy.

In the latter, the interactions among its members occurred automatically, largely free of government intervention and deliberate control, although only Great Britain practiced free trade. Continuous adjustments took place regardless of their adverse effects on each country's rate of economic growth, level of employment, pattern of production, distribution of income, and standard of living. Today, in contrast, Atlantic societies are no longer willing to endure periodic depressions, mass unemployment, massive loss of income, widespread bankruptcies of noncompetitive firms and farms, and the unchecked decline of older industries, districts and towns. Instead, Atlantic governments seek individually and in concert to manage the adjustment process so as to prevent it from affecting national economic welfare in ways no longer acceptable to their opinion leaders and their people generally. Essentially, this has meant supplementing and guiding the harmonization of national economic conditions effected through market forces by means of deliberate coordination of national economic policies among the governments concerned. The purposes of coordination are to try to prevent unacceptable domestic adjustments and to share equitably the burden of those that are tolerable.

Although such voluntary intergovernmental cooperation has helped importantly to prevent 19th-century kinds of effects, the process of continuous mutual adjustment is nevertheless sufficiently severe at both *micro* and *macro* levels to generate deeply ambivalent reactions. Avid for the benefits of regional economic integration, the Atlantic nations are unwilling to incur its costs. In essence, the momentum of the economic growth process is confronted by the inertia of institutions— their natural resistance to change in their accustomed patterns of internal and external relationships and in their familiar operating procedures. The benefits of economic growth for the majority are weighed against the harm it does to the minority of organizations and individuals unable or unwilling to adapt to changing conditions. Thus, at the *micro* level, the crux of the contemporary ambivalence about regional integration is the inability or unwillingness of business firms, workers and farmers to make painful adjustments or to accept the undesired consequences of not making them. This is reflected and paralleled at the *macro* level by the reluctance of national governments to impair their economic sovereignty—their unilateral ability to influence their national economic conditions and welfare. For, the greater the degree of economic integration, the smaller the freedom of action that participating governments have to respond to the pressures of their people to increase income gains or to prevent income losses.

At bottom, therefore, the contemporary difficulty of maintaining and extending regional economic integration reduces itself to the private and governmental resistances to the international transfers of income entailed by the process of continuous mutual adjustments within and among the countries of the Atlantic region. In contrast to the 19th-century system, in which the painful effects of the adjustment process were fatalistically or moralistically accepted, the participants in the contemporary system both want to and believe that they can restrict, offset or prevent its unfavorable consequences. But, efforts to maximize the benefits or to minimize the costs of economic integration inevitably encounter the limitation imposed by the fact that the contemporary system lacks an authority capable of deciding upon and enforcing an equitable distribution of income among the countries concerned and the necessary measures of coordinated policy. This is why, difficult as it is, the adjustment problem among the constituent sections and groups of a national economy is by nature much more tractable than it is among the nations comprising a regional economic system.

However, the interest of the Atlantic countries in preserving economic integration and the similarity of their institutions, values and behavioral norms have been sufficiently great to make possible a minimum necessary degree of voluntary cooperation and coordination of policy. Moreover, thanks to the structural and institutional changes sketched above, the Atlantic regional system has a more organic type of integration than that conferred on the 19th-century system by trade in raw materials and finished products and by portfolio investment. The ties that bind today are more deeply rooted in the organization and functioning of the constituent national economies than were those of the past. This is why the European Community must be regarded not only as an entity in itself but also as an integral part of the Atlantic regional system, even though its ambivalence about participating in Atlantic economic integration is intensified by the latter's organic character.

In sum, regional integration reflects the crucial importance to Atlantic countries of continued economic growth, which alone creates the increasing resources needed to meet their new and expanding national goals. Economic growth is fostered by, and in turn further stimulates, regional economic integration, whose effectiveness depends upon allowing much wider scope for market forces to generate rising productivity and output. In this way, regional integration helps to ease many national problems requiring increased resources. However, the greater freedom and broader compass of market forces in the regional system also exacerbate the strains and imbalances arising from domestic conflicts over goals and resource allocations and the competition among national economic systems. The resulting difficulties of mutual adjustment among Atlantic countries at both *micro* and *macro* levels in one way or another repeatedly threaten the regional system with disintegration—that is, with reactions that could deteriorate into trade wars and "beggar-thy-neighbor" policies, which would curtail growth in most, if not all, of its members.

These problems of adjustment, in turn, reinforce the pressures on national governments. On the one hand, the complex difficulties with which they must cope compel them to increase their control over their national economies despite the limitations on the exercise of their economic powers that they have accepted for the sake of regional integration. On the other hand, they are obliged to impose further restrictions on their own freedom of action by their cooperative efforts to foster the beneficial effects and offset the unacceptable consequences of the greater freedom and scope of market forces in the regional system. The inadequacies of national governments' attempts to reconcile these exceedingly complex and contradictory pressures and constraints aggravate the disintegrative threats to the system. And, because of the more organic nature of the integration already developed by the Atlantic region, the process of disintegration would today be more painful and costly and its consequences much less acceptable to social values and popular expectations than in the past.

Economic Problems and Prospects

The difficulties of national and regional policy making arising from these fundamental ambivalences have hitherto been held within tolerable limits by the fact that the positive interest of the Atlantic countries in obtaining the benefits of regional integration and, conversely, their negative interest in avoiding the costs of disintegration have constituted such powerful institutionalized factors that the regional system has tended to be self-perpetuating. Indeed, it has been self-reinforcing. As explained in the preceding chapter, all other things being equal, economic integration functions through increasing trade and investment and the other aspects of the expanding transnational integration of production to generate private and official pressures both for the autonomous harmonization of national economic conditions by market forces and for the deliberate coordination of economic policies by national governments. As in the case of European economic integration, the fate of Atlantic economic integration can most meaningfully be assessed by ascertaining whether and how the disintegrative threats to it at *micro* and *macro* levels are likely to be met.

Reactions to Increased Competition

The major disintegrative force at the *micro* level arises from the responses of business firms, workers and farmers to the increased competition in an integrated regional system. Their differing competitive capabilities are, in essence, the long-familiar comparative advantages without which there would be no gains from trade among nations, and hence no interest in incurring its costs and risks. The crux of the problem is that, while comparative advantages make trade possible among nations, the corresponding comparative disadvantages inhibit it. Agricultural and industrial producers with significant comparative disadvantages naturally seek governmental assistance to protect their domestic markets from import competition and to promote their exports in foreign markets. From time to time, governments impose restrictions on imports and subsidize exports in response to private pressures not only because such special interests are politically influential but also in the belief that establishing or preserving domestic production of the particular commodities involved is in the national interest for defense, economic or political reasons. The United States is as prone to succumb to protective pressures as are the other Atlantic nations.

These long-familiar protectionist reactions are an endemic threat to regional integration and have been responsible for the reimposition of certain trade barriers, as in textiles. Nonetheless, protectionist pressures of this type are likely to impair regional integration seriously only when they are infused with an intensified sense of importance by the injection of other interests and attitudes. In Europe during the 1960s, for example, they were magnified by the anxieties over technological, managerial and educational "gaps" and, as also in Canada, by fear of economic domination by the United States. Conversely, in the United States at the end of the decade, they were aggravated by disappointment of the expectations regarding the behavior of the European Community, as well as of Japan—a reaction which enlisted for protectionist proposals much wider support among opinion leaders and the public generally than only the members of the interest groups concerned.

Trade unions in the United States, like the farm organizations before them, have raised the question of equity regarding the effects of the adjustment process at the *micro* level: why should workers who lose their jobs or suffer reductions of income as a result of import competition or the decline of exports make the painful changes required while the other groups, as consumers, benefit without cost to themselves from the gains from international trade? And, just as in analogous circumstances of major income and employment inequities, the farmers obtained compensatory governmental aid during the interwar period, so the principle that workers were entitled to adjustment assistance was recognized in the Trade Expansion Act of 1962, which authorized U.S. participation in the Kennedy Round of tariff reductions. However, the criteria for dispensing aid were drawn and administered so strictly that none was granted until 1969, when a somewhat more liberal interpretation was instituted. Meantime, the labor movement had become disillusioned with adjustment assistance and was pressing instead for protectionist measures, such as quotas on imports and restrictions on foreign investments by American corporations that might result in loss of domestic employment—the so-called "export of jobs."

Owing to the political importance of trade unions, some import restrictions are bound to be reimposed from time to time in response to such pressures in the United States and in other Atlantic countries as well. However, the major means for dealing with the equity issue is likely to be increasing liberalization of the terms of, and the conditions and procedures for obtaining, adjustment assistance of various kinds (such as, unemployment benefits equal to previous wages, relocation allowances, retraining programs). Such an approach reflects the expanding conception of the society's welfare obligations and the broadening responsibilities of national governments in the period of the new nationalism.

In Western Europe, adjustment assistance has been much more readily available to ease the effects of increased competition. As explained in Chapter II, social-welfare programs were adopted much earlier than in the United States and have generally been broader in scope. Other types of governmental measures, such as development programs for stagnant or declining regions within countries, existed to relieve chronic unemployment and provide generous retraining and relocation benefits. Also, the high growth rates prevalent throughout the 1950s and '60s led to persistent labor shortages, thereby greatly facilitating the task of finding new jobs elsewhere in the economy.

The need in Western Europe has rather been to assuage the anxieties aroused by the competitive pressures from American multinational enterprises in the integrated regional system. And, relief has come from the very source that generated the anxiety—the stimulus to European growth and competitiveness resulting from the transnational integration of production. Since the mid-1960s, European business firms have been reacting positively to increased American competition by adaptations of U.S. management methods, growing professionalization, greater mobility of younger executive and technical personnel, larger research and development expenditures, and more aggressive marketing activities.[5] Neither the speed nor the intensity of this response would have been as great had the relationship between North America and Western Europe consisted predominantly of the 19th-century type of trade and investment.

In turn, the economic influences exerted by contemporary relationships are dependent on the sociocultural factors involved. Without the United States, the Atlantic system would lack its most important source of dynamic redemptive activism and self-confident technocratic positivism, the characteristics that help to motivate the significantly greater American willingness to improve, innovate, compete and adapt to changing circumstances. Although they differ only in degree from those of Americans, the attitudes and practices in these respects of the various European societies would have been too similar among themselves to have stimulated nearly as much or as rapid change as has their direct exposure to the behavior of American companies. This is true not only in competitive interactions but also in those of a cooperative nature. For, these greater American propensities also help to produce the financial, technical and managerial benefits that, despite their fears of American domination and takeovers, have already markedly inclined European business executives toward joint ventures, licensing agreements, and other types of cooperative arrangements with U.S. companies. Regardless of similar worries, European governments, too, have more and more recognized the important role played by direct competitive and cooperative relationships in stimulating the dynamism of European economies.

Hence, by the early 1970s, European anxieties about the presumed danger of U.S. domination through private American investment were being shifted to the threat of the disproportionate power of the multinational enterprise *per se,* regardless of the nationality of its parent company. In part, this concern expresses the anomaly that, whereas governments' own freedom of action is narrowing in consequence of the pressures of the new nationalism and the constraints of regional economic integration, that of multinational enterprises is widening as a result of the transnational integration of production. In part, too, it reflects the fact that, due to the changes in values and expectations implicit in the new nationalism, more and more of the specific ways in which business firms interact with the societies where they operate are becoming explicit objects of public policy, as explained in Chapter II.

For these reasons, growing attention is being devoted in the Atlantic countries to proposals for national and international control of multinational enterprises. National government regulation of all forms of private economic activity, and not only of multinational companies, has been increasing for decades in consequence of the sociocultural changes of the 20th century, and this trend is likely to continue. Atlantic governments may sooner or later have to harmonize those types of national regulations that significantly affect the conditions of competition in the integrated regional system. It is less likely, however, that an international authority would be established to regulate multinational enterprises on a regional or global basis owing to the inability and unwillingness of Atlantic governments in the period of the new nationalism to transfer the requisite powers to it.

In sum, because of the effects of the transnational integration of production, leads and lags in innovation and productivity will continuously shift back and forth among the Atlantic countries and between them and Japan. Thus, *le défi américain* of the late 1960s has already been converted into *le défi européen* of the early 1970s, and both are facing *le défi japonais*. In the years to come, one part of the region and then another will forge temporarily ahead in the ceaseless redistribution of comparative advantages. Protectionist reactions to *micro* competition are inherent in such an integrated system but are not likely in themselves to cause it to disintegrate.

Agricultural Protectionism

While, in the main, regional competition among industrial firms and workers results in continuous mutual adjustments eased by governmental assistance of various kinds, similar responses have not been and are not likely to become the rule in agriculture. For national defense and as part of their lingering patrimonial heritage of mercantilist policies, most continental countries have always sought to maintain a high degree of self-sufficiency in essential foodstuffs even at the cost of high prices relative to urban family incomes. The United Kingdom, too, has been moving in this direction, World War II having revealed the dangers of its heavy dependence on vulnerable seaborne imports. Moreover, except in Britain, farmers still constitute a sizable proportion of the voting population, electing a bloc of representatives whose support is often decisively important for cabinet formation. The farm population in the United States began to decline in numbers earlier and more rapidly than in most of Western Europe. Nevertheless, it remained large enough and continued to retain a sufficiently disproportionate share of Congressional seats to be a major political force until the late 1960s.

For these reasons, protectionism rather than trade liberalization has been the dominant characteristic of national agricultural policies in the Atlantic region. The Kennedy Round left tariffs on competitive agricultural products virtually untouched, and the EC's enlargement and common agricultural policy have increased the degree of protectionism in Atlantic trade in these commodities. On grains, dairy products, meat, poultry, fruit and other commodities grown in substantial amounts in EC member nations, the common agricultural policy requires that a variable levy be imposed, equal to the difference between the world market price and the support price within the Community. Since they were initially set in 1965, support prices have been repeatedly raised in response to pressure from the farmers seeking to offset the effects of inflation and to improve their real incomes. In turn, high support prices stimulated increased production, leading to reduced imports and the accumulation of burdensome surpluses of certain commodities. Under the common agricultural policy, revenues obtained from tariffs levied on agricultural imports, supplemented by contributions from national budgets, are used to acquire unmarketable surpluses and to subsidize their export, which from time to time has resulted in serious loss of sales by the normal exporters of these products.

Like most other countries assisting their farmers, the United States has used a system involving high support prices and the accumulation of surpluses by the government. However, owing to the greater productivity of American agriculture, U.S. support prices are generally lower than those of the Community. The United States also imposes barriers to imports of supported commodities, and in effect subsidizes the export of surpluses through its food-aid programs.

Although U.S. agricultural policy differs only in degree from that of the Community, there are important conflicts of interest between them which have led, and are likely to continue to lead, to serious transatlantic disputes. While total U.S. exports to the Community have been growing, the agricultural portion has lagged behind. Exports of commodities covered by variable levies declined substantially after their high point in the mid–1960s. It was not until 1971 that the increase of exports of products, such as oilseeds, not covered by variable levies raised total U.S. agricultural exports to the Community above the 1966 level. Also, the United States and the other major exporters of temperate-zone agricultural products object to the adverse effects on the world market of the EC's subsidized exports of surplus commodities.

Both in Europe and in the United States, policy makers generally envisage that the declining size of the farm population in Atlantic countries will sooner or later permit a more rational approach to the problems of agricultural production, incomes and regional trade. The political importance of the rural vote and the pressures for governmental assistance are expected to diminish as many less efficient farmers retire or are absorbed into other economic activities. These long-term demographic and economic changes could be eased and accelerated, it is believed, by shifting from a price-support to an income-support system of government assistance, by early retirement pensions and relocation allowances, and by various measures for fostering more efficient use of agricultural land, labor and capital. In Europe, the EC Commission has sponsored a comprehensive program along these lines, known as the Mansholt Plan, but as yet only a few minor measures for encouraging improvements in productivity and the movement to other economic sectors have been approved by member governments. In the United States, the declining importance of the farm bloc following Congressional redistricting in the late 1960s has enabled a beginning to be made in reducing some high support prices and replacing the lost benefits by "deficiency payments" to low-income farmers.

These first tentative moves provide some grounds for the expectation that, in the course of the 1970s, enough inefficient farmers will be eliminated from the agricultural sectors of the Atlantic nations to allow substantial reduction, if not complete removal, of price supports. This development, it is hoped, would permit liberalization of regional trade, with agreement to impose production controls to prevent accumulation of unneeded surpluses. In such circumstances of freer trade, the comparative advantages of soil and terrain enjoyed by North American farmers would enable them at least to preserve and perhaps to increase their share of the European market for temperate-zone agricultural products.

However, there is also evidence for a contrary outcome. As the remaining European farmers become more and more efficient through enlargement of cultivating units, greater capital investment, and improved inputs and methods, their productive capacity will *ipso facto* increase, and they will be impelled to offset the effects of lower prices on their incomes by raising and marketing larger amounts of the crops involved. This reaction has already occurred in the United States and has led to the imposition of production controls. But, efficient European farmers may be better able to resist production controls because their political importance may not decline as much or as rapidly as that of American farmers, and their efforts may be reinforced by the long-standing commitment of their governments to high degrees of national self-sufficiency. So far, too, European farmers have refused to accept and European politicians have been unwilling to impose any significant substitution of income-support for price-support methods. Both prefer the latter technique, which distributes the largest part of the cost among consumers generally rather than making it a conspicuous and vulnerable item in the national budget.

For these reasons, it is by no means certain that long-term demographic and economic changes will lead to substantially lower prices or that, if they do, the increasing productive potential in Europe will be sufficiently inhibited by lack of incentives or by production controls to permit the liberalization of regional trade in temperate-zone agricultural commodities. The factors operating for and against liberalization are closely enough balanced in the shorter term to preclude a reasonable assessment of probabilities at this time. The crucial decisions are likely to be those made by the European Community regarding the adoption and implementation of the Mansholt Plan or its equivalent. If, in the next two or three years, the essential measures relating to price reductions, income supports, and the elimination of inefficient farmers are accepted and put vigorously into effect, the chances for eventual liberalization will steadily improve. If not, regional agricultural trade will be increasingly restricted as European production rises, and transatlantic conflicts over the issues involved will worsen. In turn, these disputes could sufficiently reinforce the protectionist pressures from business firms and workers in the United States to make the *micro*-competition problems of regional integration much more serious disintegrative threats.

Problems of Internal and External Balance

The continuous mutual adjustments inherent in an integrated regional economic system manifest themselves not only in competition at the *micro* level among individual producing and investing units but also in imbalances at the *macro* level within and among the constituent national economies as a whole. Like the difficulties of competition, the problems of imbalance are inseparable from the benefits of regional integration yet continually threaten to disintegrate the system. And, even more than the strains arising from *micro* competition, they are raised to critical intensity by the changes in institutions and values characteristic of the period of the new nationalism. The disintegrative pressures at the *macro* level typically take two interrelated forms. The first expresses the difficulty of preserving sufficient freedom of action in domestic economic affairs to maintain reasonable price stability and employment while realizing elite-group and popular expectations regarding expanding national goals. The second reflects the difficulty in these circumstances of maintaining reasonable external balance of payments under the conditions imposed by international monetary arrangements.

In the dynamic affluent societies of the Atlantic countries, the increasing diversity of national, group and individual goals, the rising intensity with which they are pursued, and the consequent growing competition for resources are expressed in greater politicizing of the process of resource allocation. Economic and political decisions and compromises affecting resource allocations are rarely once and for all choices; they must continually be made anew. So persistently do claims proliferate and so insistent are the demands generated by the internal crises resulting from rapid social change and by the insecurity of the international political system that resource allocations, explicit and implicit, generally exceed resource availabilities. In consequence, the competition for resources is manifested not only in domestic conflicts but also as a recurrent difficulty of balancing growing supply against even more rapidly rising demand—in short, as the problem of inflation. And, neither on the supply side nor on the demand side is there an effective prescription for keeping a modern Atlantic economy in reasonable balance.

Economic growth increases the availability of resources for meeting the claims on them. If this were its sole result, economic growth would, indeed, be the sovereign remedy for inflation. But, the growth rate cannot be pushed higher than a certain magnitude—which varies from country to country—without itself engendering inflationary pressures. As an economy operates closer and closer to its full productive capacity, not only is total demand *ipso facto* expanded, but inflationary pressures are further intensified by the competitive bidding also stimulated for increasingly scarce factors of production—such as skilled labor, capital equipment, materials and parts. And, in the period of the new nationalism, pressures for upward wage and price adjustments persist even after disinflationary measures have reduced employment of labor and facilities—the so-called "cost-push" phase of inflation.

If the limitations on the supply side are largely imposed by the nature of the economic process itself, those on the demand side arise from contemporary sociocultural changes and political and international constraints. The commitment to full employment and to at least the maintenance of existing living standards makes it very difficult for any national administration to restore or preserve internal balance by rigorous deflationary means. For such internationally active nations as the United States—and France under de Gaulle—the pressures to eliminate poverty, improve education, advance basic and applied knowledge, protect the environment, and achieve other social goals have made it equally difficult to reduce resource allocations for these purposes in order to offset the mounting costs of defense and foreign-policy expenditures. In these and others ways, the commitment to new and expanding national goals not only helps to generate inflation but also imposes fairly narrow limits on the kinds of measures that can be taken to cope with rising prices by restraining demand.

For these reasons, *the tendency to inflation is now, and will continue to be, endemic in the Atlantic region.* Although its incidence will vary, few, if any, members of the Atlantic system are likely to be permanently free of the difficulties engendered by their own or other nations' inflationary pressures. As they have since the new period began, most Atlantic countries will continue to oscillate over a greater or lesser range between the extremes of inflation and deflation. At one time, they will push up or allow the growth rate to rise to the point of overheating the economy, with attendant sociopolitical problems and often serious balance-of-payments difficulties, and, at another time, they will reduce or hold down the growth rate, with concomitant slowing or even loss of social-welfare gains that sooner or later will generate severe social tensions and domestic political conflicts.

The problem of internal balance interacts with the difficulty of maintaining reasonable equilibrium in the nation's external accounts. Because it expresses a relationship between a national economy and its major trading and investing partners, a disequilibrium in a country's balance of payments is never solely the result either of internal or of external factors.[6] In the 19th-century system, these interrelationships across national boundaries were largely self-equilibrating. Changes in comparative prices and interest rates stimulated changes in trade and capital movements, and *vice versa,* that rapidly, and sometimes very drastically, deflated countries with substantial deficits and inflated those with substantial surpluses relative to one another. However, this self-equilibrating mechanism periodically permitted the familiar commercial crises and money panics which the 19th-century system accepted as inescapable and the contemporary Atlantic system rejects as unnecessary. Hence, the interrelationship of internal and external factors in payments imbalances has ceased to be only a truism of economics and has become a major concern of national policy.

The internal measures for dealing with serious balance-of-payments deficits are similar to those for coping with excessive domestic inflationary pressures and hence have the same limitations. Too high a rate of economic growth can stimulate disproportionate increases in the demand for imports, while limiting or even reducing export capabilities. The same effects can result from over-allocation of resources to achieve pressing national goals, and the ensuing adverse consequences for the balance of payments can be magnified by mounting outlays for international-security and foreign-policy objectives. Deflationary policies designed to limit or relieve such balance-of-payments strains cannot cut very deeply or quickly if they are to be socially acceptable.

External measures for dealing with a severe payments imbalance have limitations of their own. Memories of the failures of the interwar period reinforce the pressures to preserve regional integration in restricting the use of neomercantilist devices to critical situations. Within this self-imposed limit, countries with serious imbalances have from time to time resorted to import surcharges and quotas, export subsidies of various kinds, and controls on private capital movements. Considering, however, the persistence since the late 1950s of anxiety-generating payments imbalances and the related recurrent international monetary crises, the use of restrictionist measures has so far been minor, where permanent, and temporary, where substantial (as in the cases of the U.K. and U.S. import surcharges). This restraint reflects the determination of Atlantic countries not to regress into neomercantilism.

The ability to resist restrictionism depends upon two other means for coping with payments imbalances. The first consists of the resources available to a country for financing a balance-of-payments deficit—essentially, its own reserve assets and the credits it can obtain from other countries and the IMF. The second is the ability to alter the exchange rate of its currency so as to appreciate or depreciate its value *vis-à-vis* the currencies of other members of the regional system.

To the extent to which a country possesses reserve assets acceptable to other nations—such as gold and convertible currencies—it has means for financing a deficit in its balance of payments. The use of reserves gives it time for the deficit to disappear or shrink to minor proportions either through anticipated autonomous changes in trade and capital movements or through those deliberately induced by policy measures. Moreover, the possession of large reserves confers on a nation the prestige still enjoyed by the rich and enhances its influence in international economic affairs. In a unique coincidence, the lingering influences of patrimonial prejudices for conspicuous wealth and of the worldly asceticism of the Protestant ethic have combined to sustain the tendency to judge the success or failure of the national monetary authorities (treasury officials and central bankers) in managing their countries' external economic relations by the size of the reserves they accumulate.

Large reserves are not always an unmixed benefit; if they are too big or grow too rapidly, they can have an inflationary effect within the economy. Even so, the factors fostering the accumulation of reserves are felt to be so compelling that it has itself become an objective of policy competing for resources with other national goals. The best known example was Gaullist France which, until the unsettling events of May 1968, managed its economy so as to have a continuing balance-of-payments surplus, thereby enabling it to increase its reserves very markedly. In greater or lesser degree, all Atlantic countries are impelled to try to augment their reserves and are reluctant to allow them to decline substantially. This universal, though variable, propensity to accumulate and hold monetary reserves has major implications not only for domestic welfare but also for the functioning of the international monetary system, as explained in the next section.

Changes in exchange rates, too, have important internal repercussions. A rise in the value of a country's currency relative to those of others has the effect of lowering the prices of its imports and raising those of its exports, thereby tending to stimulate the former and restrain the latter. Reducing the exchange rate of a currency has the opposite result—exports are encouraged while imports are inhibited. The nature and extent of the effects of exchange-rate adjustments also depend upon whether they are small or large, gradual or sudden.

In theory, such changes are supposed to be made to take account of "fundamental disequilibria"—that is, substantial and irreversible shifts in the productivity and real costs of the factors determining the competitive position of a country *vis-à-vis* its major trading rivals. In practice, however, the Atlantic nations were reluctant during the 1950s and '60s to make exchange-rate changes for several reasons. It is not always possible to determine beyond reasonable doubt that a fundamental disequilibrium has occurred. Also, memories of the harmful competitive devaluations and of the unsatisfactory experience of trying to manage rates with no fixed par values during the interwar period, as well as considerations of national prestige and of domestic partisan politics, have helped to deter reliance upon this form of external adjustment. Finally, exchange-rate modifications do not simply reflect basic changes in a national economy; if too large or too frequent, they also have consequences for employment and living standards that might be no more acceptable to a country than internal deflationary or inflationary measures directly affecting the level of demand. For their part, domestic political and economic restraints can operate to postpone, diminish or nullify the intended benefits of exchange-rate changes, thereby making them that much slower or less effective in helping to restore balance, as happened in the United Kingdom after the sterling devaluation of 1967.

Thus, the inability of Atlantic countries to keep their external imbalances at noncritical levels does not reflect irrational perversity, as public discussion sometimes implies. It arises essentially from the sociopolitical limitations explained above and, in the case of the

United States, also from defense and foreign-policy commitments it believes indispensable for its own and its allies' security. These constraints are reinforced by the absence of a sense of urgency about external payments difficulties among their people generally. Balance-of-payments problems are usually regarded as highly technical. More important, their relevance to the conditions of life tend to be perceived as much less immediate and direct—though actually they may be no less pervasive and serious—than that of the problem of internal balance, whose consequences, as an economy approaches either end of the inflation-deflation range, directly and powerfully affect the well-being of large numbers of people.

The Structural Anomalies of the International Monetary System

Balance-of-payments problems, especially of the United States, are of major significance for regional economic integration not only because of their importance for national policies but also because of their role in the international monetary system as a whole.[7] Established at the Bretton Woods Conference in 1944, this system was fully activated only at the end of 1958, when the current-account convertibility of European currencies was restored. Since then, the international monetary system has been plagued by recurrent crises that are inherent in its structural anomalies and are triggered by loss of confidence in one or more of its leading currencies, in most cases engendered by large, persisting balance-of-payments deficits.

In an international monetary crisis, the governmental and private holders of monetary assets seek to convert them into forms believed less likely to decline in value or, especially in the case of speculators, more likely to increase in value. True, the ordinary mechanism for financing international trade and investment involves sizable daily movements of funds among the various Atlantic currencies. In a monetary crisis, however, the movement is much larger and more rapid than normal and predominantly in a single direction—that is, into the one or two currencies regarded as strongest, or out of a weak currency, or from currencies into gold. The problem of confidence, therefore, takes the form of large abrupt changes in the monetary-asset preferences of governments and private organizations, which can rapidly drain a country's reserves of internationally acceptable forms and correspondingly increase its need to obtain the additional amounts of reserve assets that are available to it in the international monetary system. The total volume of such readily available and generally acceptable reserve assets is known as the system's liquidity. Since Bretton Woods, gold and dollars have been the chief forms of reserve assets, supplemented by small amounts of sterling, marks and other currencies, and of IMF borrowing quotas and special drawing rights (SDRs).

The serious payments imbalances that often precipitate monetary crises reflect changes not only in trade flows but also in capital movements, private and governmental. At the Bretton Woods Conference, it was envisaged that, as during the interwar years, member nations would continue to restrict and control capital movements as a major means for maintaining or restoring external balance. However, both resulting from and further stimulating regional economic growth and integration, the Atlantic countries liberalized, although most did not completely abolish, their restrictions on capital movements by the late 1950s. In consequence thereafter, the flows of short- and long-term funds increased even faster than the rapid expansion of trade. Because of the importance of the benefits derived from capital movements, especially in the Eurocurrency market, most Atlantic countries have been reluctant to tighten their limitations on capital flows and credit creation to help achieve balance, and they have been additionally deterred by the difficulty of making more restrictive controls effective.

Capital movements are inherently more fluid than trade movements, responding more readily to possible gains arising from disparities in economic conditions and prospects, especially interest differentials, and to exchange-rate risks. This greater susceptibility of capital movements—both of liquid (short-term) capital and of new funds seeking long-term investments—to changes in volume and direction is superimposed upon the volatility of money flows for commercial payments (the so-called "leads and lags") among Atlantic countries, and is made more fluid by the scale and flexibility of the Eurocurrency market. On the one hand, the magnitude and fluidity of capital movements make it possible for them to act rapidly as equilibrating factors offsetting other changes in the balance of payments, as they did in the 19th century. On the other hand, the prominence and sensitivity of capital transactions often initially exaggerate the size of payments imbalances, especially during crises of confidence in currencies, and thereby stimulate the felt need for larger reserves and credit facilities to cope with them. This desire for increased liquidity in the system has been reinforced by the fear that countries with inadequate reserves might be compelled by payments difficulties to reimpose import restrictions and capital controls detrimental to the high and growing volume of trade in the Atlantic region and the transnational integration of production.

While, since the late 1950s, the demand for internationally acceptable reserve assets has in consequence been rising, increases in the supply of gold for meeting it have been grossly insufficient. This difficulty reflects the anomaly that the factors determining the supply of gold—the rate of new ore discoveries and their gold content, the profitability of gold mining relative to the fixed monetary price of gold, the volume of Soviet sales of gold—are not by nature adequately responsive to the factors determining the demand for gold for monetary purposes, for private speculation and hoarding, and for industrial and other nonmonetary uses. Nor were IMF borrowing quotas sufficient to meet expanding liquidity needs. Although members' quotas were thrice raised—in 1959, 1966 and 1970—the additional liquidity generated by this means constituted a very small fraction of the total available. For these reasons, the "key currencies" of the interwar period—primarily the dollar and, to a much lesser extent, sterling—became the principal means for supplying the increase in the liquidity of the international monetary system after World War II. By the end of 1965, dollars constituted nearly a quarter of total world monetary reserves of $71 billion, and they rose to more than a third of $92.5 billion by the end of 1970. During the prolonged dollar crisis of 1971, European and Japanese central banks acquired such large additional amounts of dollars in their efforts to limit the appreciation of their currencies that, by the end of that year, dollars are estimated to have comprised as much as 46 percent of total world monetary reserves of $130 billion.[8]

The U.S. balance-of-payments deficit has been in effect the means for making possible most of the increase in liquidity since 1950. Unwilling to institute the drastic changes in its national policies that would be required to restore external balance, the United States has been determined to have its deficit financed by other countries. In turn, until the 1971 crisis, the latter were willing—indeed, some were eager—to accumulate dollar reserves by doing so. Thus, the U.S. deficit has been the indispensable counterpart of increased liquidity. At the same time, however, the size and persistence of the deficit led official and private holders of dollars to convert some of them into other reserve assets, resulting in the drastic decline of the U.S. gold reserve from nearly $25 billion in 1949 to slightly over $10 billion during the crisis of 1971, when the gold convertibility of the dollar had to be abandoned. In this way, the very means by which dollars have been made available to meet the demand for additional liquidity during the 1960s has caused confidence in them as international reserve assets to diminish the more they have been used for this purpose.

Just as in the case of the inadequate supply of gold to the international monetary system, the factors affecting confidence in the dollar are insufficiently related to the special function it performs. This anomaly reflects the fact that the dollar is not essentially an *international* monetary asset but the *national* currency of the United States. On the one hand, the other Atlantic nations would not be willing to accumulate dollars to so large a proportion of their monetary reserves if they did not have a great deal of confidence in the wealth and productivity of the American economy and the good faith of the U.S. government. On the other hand, they feel considerable anxiety over their inability to control U.S. government decisions affecting the international role of the dollar so as to assure adequate regard for their own interests. Moreover, the Europeans increasingly resented the fact that, by accumulating dollars in their reserves, they were implicitly assisting American policies, such as the Indochina intervention, which contributed to the U.S. deficit and which they believed to be mistaken.

These ambivalent interests regarding the structural problems of the IMF system, as well as the misconceptions about them, have been responsible for the slowness of the effort to formulate and adopt changes that could substantially reduce, if not eliminate, the propensity to recurrent crises of confidence. All members of the Group of Ten (see footnote 7), including initially the United States, were for years reluctant to recognize that the growth of liquidity in a way adequately responsive to the demand for it could only come about by means of a form of reserve asset that was neither primarily a national currency, like the dollar, nor a commodity, like gold, with other monetary and nonmonetary uses. The prototype of such an asset already existed in the IMF borrowing quotas. Hence, after years of agonizing and discussion, the decision was finally reached at the IMF's annual meeting in 1967 to develop further the potentialities of this type of asset for expanding international liquidity. By then, moreover, the members of the Group of Ten were prepared to concede that it would be preferable for this purpose to establish a new category of special drawing rights (SDRs), whose volume could be more readily increased as the need for additional liquidity was felt, which would be more freely available to member countries than the regular borrowing quotas, and which, unlike the latter, would not have to be repaid (although countries would be subject to the limitation that not more than 70 percent of their

SDRs could be outstanding over designated periods). The SDR form of international monetary asset was activated at the beginning of 1970 under an agreement to create $9.5 billion of them over the ensuing three years.

Unlike gold or dollars, the conditional IMF borrowing quotas and the automatically available SDRs are forms of monetary asset that are essentially fiduciary and international in nature. These characteristics are responsible, on the one hand, for the superior potential utility of SDRs for coping with the liquidity problem and, on the other, for their peculiar potential vulnerability. The very existence and general acceptability of SDRs are voluntary conventions—a set of beliefs and modes of behavior—and they are not maintained by the sovereign power of a national government, as is the fiduciary money of a country. (True, the acceptability even of gold is, at bottom, also a cultural convention but one that has much deeper roots both historically and psychologically.) Instead, SDRs are dependent upon a very considerable degree of mutual faith and cooperation among independent states, whose interests and attitudes are often different and sometimes conflicting. Moreover, the consequences are still largely unknown of relying increasingly upon an international fiduciary asset for the major portion of the future growth of liquidity in the IMF system. For these reasons, the governments of the Atlantic nations, and particularly their monetary authorities, have persisted in hoping that the system could continue to function without significant changes, have reluctantly agreed to innovate only under the compulsion of necessity, and then have been prepared to make only the minimum modifications required at the time.

The structural anomalies of the international monetary system are reflected not only in the liquidity and confidence problems but also in the difficulties of the *macro*-adjustment process. The objective of the latter is not, of course, to try to eliminate balance-of-payments deficits and surpluses or even to keep them very small. Neither is possible in a system of relatively free trade and capital movements. What is required for continued regional integration is that the tendency toward large prolonged deficits and surpluses, especially by leading countries of the system, be inhibited and that member nations be willing and able to adopt measures, singly and in concert, for moving toward balance, with adequate means of financing available to them during the interim and in critical situations. In other words, *macro* adjustments are always necessary; the task is to make them effective and acceptable despite the conflicts among the requirements of continued regional economic integration and the restrictions imposed by the new nationalism.

One possibility under increasing discussion since the early 1960s' for moderating the frequency and severity of monetary crises and easing the adjustment process is greater flexibility in exchange rates. Indeed, some economists advocate completely free rates determined by supply and demand as the remedy for the problems of the international monetary system. Despite its theoretical merits, however, such complete flexibility of exchange rates probably has limited applicability in real-life situations. Exchange-rate changes are not without internal economic consequences, which may be no more acceptable than direct deflationary or inflationary measures. Also, although the extent to which uncertainty about exchange rates discourages international trade and investment is exaggerated by the opponents of free rates, such effects do occur and could be quite significant over the long term for countries whose currencies fluctuated too widely and frequently.

These objections do not apply in anywhere near the same degree to the various proposals for limited—instead of complete—flexibility in exchange rates. One way to obtain such limited flexibility is to widen the band within which exchange rates are permitted to fluctuate under IMF rules; another would be to add to this change an arrangement— usually called the "crawling peg"—for enabling par values to move by small amounts over specified periods either automatically as determined by the market or in accordance with an agreed-upon formula. However, to mitigate the adjustment problem substantially, limited flexibility would have to be combined with a reversal of the basic attitude regarding deliberate changes in official par values that has prevailed since World War II.

Although the Bretton Woods Conference envisaged regular use of exchange-rate adjustments, a change in an official par value soon came to be regarded as a measure of last resort, for reasons explained earlier. However, the beneficial effects of the devaluation of sterling and the French franc and of the revaluation of the German mark in the late 1960s and early 1970s began to foster a more favorable attitude toward exchange-rate changes. This shift was confirmed by the devaluation of the dollar and the revaluation of the Japanese yen, the mark and other European currencies after the dollar crisis of 1971. At the same time, the Group of Ten also agreed to increase the width of the band within which their exchange rates could fluctuate from 1 percent on either side of par value to 2¼ percent.

A similarly slow development has characterized the effort to ease the *macro*-adjustment process through more effective coordination of national fiscal and monetary policies. The objective is to narrow the divergence in size and timing of inflationary and deflationary trends within Atlantic countries, especially in the system's leading members, and thereby to prevent or reduce disruptively large movements of goods and capital across national boundaries. Because of the immediate and powerful effects of interest-rate differentials on capital movements, monetary policy has naturally attracted the most attention, especially in the monthly central bankers' meetings at the Bank for International Settlements (BIS) in Basle. Useful discussions are also conducted in the OECD, where the Atlantic countries regularly review each others' national economic conditions and policies and make suggestions for changes in them to cope more effectively with excessive balance-of-payments surpluses and deficits.

Although *macro*-coordinating activities that go beyond the immediate needs of crisis management are likely to continue and to grow, they, too, are inhibited by the limitations imposed on Atlantic governments by their problems of resource allocation and internal balance—and additionally in the case of the United States by the world role it feels impelled to play. These constraints make most Atlantic nation-states unwilling to accept explicit formalized coordinating commitments restricting their freedom of action, even though they are willing in practice to do so implicitly to an unprecedented degree. Hence, it is likely that, as with other means of preserving and improving economic integration, the deliberate coordination of national economic policies will develop quite slowly—by minimum steps under the pressure of necessity.

The net result of these opposing pressures will probably be increasing, though reluctant, recognition by Atlantic countries that the *macro*-adjustment problem cannot be eliminated, even if its magnitude can be significantly reduced by more frequent changes in par values, greater flexibility of exchange rates, and better coordination of national economic policies. Accordingly, if regional economic integration is to be preserved, much less intensified, the Atlantic system will have to live with persisting payments imbalances. Means for helping to finance such imbalances are available both in the IMF's borrowing quotas and SDRs and in the bilateral and multilateral arrangements for short-term emergency credits among the Group of Ten.

The only payments imbalance that could threaten to disrupt the system would be excessively large and prolonged deficits on the part of the United States. Even if, in a severe monetary crisis, another leading nation, such as the United Kingdom or France, could not obtain adequate financial assistance and were forced to reimpose substantial trade and payments restrictions—in effect, withdrawing temporarily from the European Community and the Atlantic economic system— neither form of economic integration need be fatally impaired. Only if the United States were unable to finance a deficit—in this event, unmanageable by definition—would the danger of serious disintegration arise.

The 1971 devaluation of the dollar relative both to other leading currencies and to gold undoubtedly restored some, if not all, of the price competitiveness that the United States had lost owing to the substantial increases in European and Japanese productivity during the 1950s and '60s. But, as the British devaluation of 1967 showed, the beneficial effects of such relative changes may take as much as two years to manifest themselves fully. Hence, at present writing, it is too early to tell whether the U.S. balance-of-payments problem will be reduced to manageable size and, more important, how long the improvement will last. Both questions depend essentially upon the relative changes in productivity and in degree of inflationary pressure in the United States compared with other major trading nations in the Atlantic regional system, including Japan. Such complex changes and interactions are exceedingly difficult to forecast for any considerable period ahead.[9]

If serious U.S. deficits recur, they could be corrected by further changes in the exchange rate of the dollar relative to those of other major trading nations. But, so long as dollars constitute a substantial portion of world monetary assets, these countries will be unwilling to incur the loss in the value of their reserves resulting from too-frequent dollar devaluations. Should large uncorrected U.S. deficits persist, therefore, the structural anomalies of the international monetary system will continue to make it susceptible to crises of confidence despite the easing of the liquidity problem through increasing the volume of SDRs and the mitigation of the adjustment problem through widening the bands for exchange-rate fluctuations. Although very useful, these reforms palliate the structural anomalies, they do not remove them.

Indeed, the propensity of the international monetary system to generate serious issues between the United States, on the one hand, and the European Community and Japan, on the other, has been significantly increased by the former's inability since the 1971 crisis to make the dollar convertible into gold at the request of the latter, as it was during the 1950s and '60s. In consequence, the other Atlantic nations have even less freedom of choice regarding the international role of the dollar than they did before, and the ambivalences of the Europeans and the Japanese are correspondingly deepened. For these reasons, international monetary problems are likely to continue to be major threats to Atlantic economic integration in the foreseeable future.

The Tripolarization of the Atlantic Economic System

The efforts of Atlantic countries to deal with the problems and pressures of *micro* competition and *macro* adjustment will interact with the conflicting constraints of regional integration and the new nationalism in shaping the future of the Atlantic economic system. In retrospect, it can be seen that these complex interactions have manifested themselves in certain long-term trends that have characterized the evolution of the regional system. How are they likely to develop in the years to come?

One trend can be discerned in the evolution of the international monetary system in the course of the 20th century. This trend has had two aspects. The first is a gradual easing of the liquidity problem through supplementing the initial commodity forms of internationally acceptable monetary assets (i.e., gold and silver) with varieties of fiduciary forms (i.e., sterling, dollars, IMF drawing rights). The second is the mitigation of the adjustment problem through the adoption of agreed-upon rules of proper international monetary behavior, the provision of emergency credits of various kinds, the establishment of an institution, the IMF, to help apply the regulations and administer the financial assistance, and the greater flexibility of exchange rates. These interrelated changes in the international monetary system are similar to, but lag considerably behind, monetary developments within national economies, that is, the gradual reduction and eventual elimination of commodity money (except as small change), and the adoption of appropriate institutional means (such as central banks and commercial banks) and regulatory measures to ensure that domestic needs for money and credit are met and to help cope with the problems of maintaining internal and external balance. Hence, the next logical evolutionary steps for the international monetary system would appear to be to replace both gold and dollars in monetary reserves by a truly international fiduciary asset, such as SDRs, and to increase the scope and effectiveness of the international regulation of money and credit.

Nor have proposals been lacking for converting the IMF (or the BIS, or a new institution established for the purpose) into the equivalent of a regional central bank to achieve these next steps. Agreement to handle the problem of reserve-asset preferences by requiring member countries to deposit their gold and dollar reserves with the IMF; or for it to fund on a long-term basis the unwanted accumulations of dollars in national reserves; or to grant it effective authority to compel chronic deficit and surplus nations to modify their exchange rates and adopt other policies for dealing with their payments imbalances, would be in effect, if not necessarily in announced intent, significant advances toward central-bank development. More far-reaching changes would empower the IMF to issue and administer a common currency for national reserves and intergovernmental transactions in the Atlantic region, and to make investments in member countries at its own discretion, including "open market" operations, which would be a potent means for harmonizing national economic conditions.

Such measures would unquestionably be major logical developments in the long-term evolution of the Atlantic monetary system. When completed, they would result in the existence of a single fiduciary international monetary asset based on a regionalized debt, a lender of last resort with unlimited liquidity to stop or prevent crises of confidence, and a supranational authority to harmonize national economic conditions by means of its discretionary power to control interest rates and credit availabilities in the member economies. The operations of such a regional central bank would, in turn, transform Atlantic economic integration into political unification. For, it would convert the problems of *macro* adjustment from the inherently less tractable kind that occur among sovereign nation-states into the much more manageable type that exist within a single politicoeconomic jurisdiction.

This qualitative transformation, however, points to the heart of the difficulty of such far-reaching monetary evolution in the Atlantic region. The greater efficiency of a regional central bank in preserving and increasing the benefits of Atlantic economic integration could be obtained only at the sacrifice of national economic and political sovereignty that the Atlantic countries are unlikely to make in the period of the new nationalism. Indeed, it is as yet far from certain that the European nations would be willing to make such a sacrifice for the sake of European union, a much more accessible and congenial goal for them than Atlantic unification. Nor would the United States be inclined to impair fundamentally its existing degree of freedom of action in domestic and international affairs. Hence, substantial progress in the foreseeable future toward an Atlantic monetary union has a much lower probability than that toward a European monetary union analyzed in the preceding chapter.

The fact that such Atlantic monetary unification has a very low probability does not mean that nothing can or is likely to be done to improve the functioning of the international monetary system and to ease the difficulties of the *macro*-adjustment process. But, the adoption of the kinds of monetary and related measures that could do so and that would have a substantially higher probability depends upon the future development of the second basic long-term trend operating in the Atlantic economic system. This is the trend toward increasing liberalization of trade among Atlantic countries through the progressive reduction or abolition of tariffs and quotas and the harmonization of so-called "nontariff barriers and distortions."

Continuation of this second trend is fostered by the logic of economic integration which, *ceteris paribus,* works toward the lowering of trade barriers and the equalization of the conditions of competition. Except for agricultural commodities generally and the various chronically noncompetitive industrial products in Atlantic countries, tariffs are now low enough for another round of successful GATT negotiations to bring the region as a whole to the threshold of free trade in manufactured goods.

Unlike the monetary developments leading to the establishment of a regional central bank, the movement toward freer regional trade, and its maintenance, would not necessitate creation of a supranational authority, as the success of both the EC's customs union and the EFTA demonstrates. On this ground alone, it would have a higher probability than Atlantic monetary unification. Under a free-trade arrangement, not only would economically sounder means than tariffs and quotas be available for dealing in a coordinated manner with the regional problems of *micro* competition. Greater scope would also be given to market forces for gradually narrowing the disparities in national economic conditions that help to make the regional problems of *macro* adjustment so difficult to resolve.

Further reduction of trade barriers in the course of the 1970s has for these reasons a much better chance than monetary unification. However, it is less probable that it would go as far as, or take the explicit form of, a free-trade area or customs union for the Atlantic region as a whole. Neither the United States and Canada, nor the Community and its European associates, nor Japan and the other Pacific members are likely to be willing to abolish all tariffs and quotas, including those affecting their protected agricultural and noncompetitive industrial products. Equally important, the Community will continue to value its common external tariff not simply for *micro-*protectionist purposes but also as a major means for preserving its own integration and sense of identity until other ties develop that can do so more effectively. Also, the Community and the other West European nations will be very reluctant to forgo the preferential advantages they derive from their discriminatory association agreements even for the sake of an Atlantic-wide free-trade arrangement.

What form, then, would further development of the long-term liberalization trend be likely to take? Here, a third basic trend within the Atlantic system points to the probable answer. Evident since the beginning of the 1960s, this is the progressive narrowing of the economic disparities between the United States and the other Atlantic nations. In part, this development reflects the persisting high rates of economic growth in most European countries and Japan. In part, it expresses the effects of European economic integration, which is impelling the Community's members more and more to think and act as a single entity in their economic relations with the United States and Japan. These changes reinforce the pressures of the new nationalism in fostering the willingness and ability of the Community and Japan to seek greater scope for independent action in economic affairs and to deal with the United States on a more equal basis. Conversely, the relative economic power of the United States *vis-à-vis* the Community and Japan is declining and its capacity to influence their economic behavior is correspondingly diminished. The United States can no longer pressure the Community and Japan to take actions against their will except by forcible measures, as it did by unilaterally imposing an import surcharge during the dollar crisis of 1971 in order to compel them to revalue their currencies.

In consequence of the third basic trend, the structure of the Atlantic economic system is now being transformed. Throughout the postwar period, it was organized in a unipolar manner—the United States constituted its overwhelmingly powerful resource and decision center with each of the other members individually oriented toward it. Today, the system is fast becoming tripolar, with three large resource and decision centers: the United States, the European Community and Japan. None is any longer, or would be likely in the future to become, sufficiently disproportionate in economic power to dominate the others. Hence, the interactions among these three foci of the Atlantic economic system will largely determine the kind and degree of economic integration it will have in the decades ahead. In other words, the difficulties at *micro* and *macro* levels will be resolved—or alternatively, could disrupt the system—within the framework set by the emerging tripolar structure of the regional economy.

The most likely course of development, therefore, is that economic integration would proceed on a differential basis in the Atlantic economic system, as well as between it and countries elsewhere in the world. This means that both the Atlantic regional economy and the international economy as a whole would be moving away from, rather than toward, the goal of universal, multilateral, nondiscriminatory trade based on the most-favored-nation principle that has been the official objective of U.S. foreign economic policy since the days of Cordell Hull. Instead, the regional and the world systems would be characterized by groupings of countries—not necessarily contiguous geographically—of varying size and degree of preferential treatment. Today, the European Community and its associated countries in Europe, West Asia and Africa already constitute one such world trade and investment bloc; and the Soviet Union, the East European states, and the other Soviet client nations form another. In the course of the 1970s, additional economic blocs centered on the United States, Japan and China are likely to emerge, with associated countries not only in their own regions but also in other parts of the world. As for the remaining smaller members of the Atlantic system, Canada's existing extensive economic integration with the United States leaves it little choice with respect to its future major orientation, despite its fear of American domination and concern to foster its sense of national identity. Australia and New Zealand, however, could gravitate toward the Community, the United States or Japan, depending on which set of trade and investment relationships becomes more important to them in the next five or six years.

In this projection, further reductions of trade barriers among the European Community, Japan and the United States would tend to be less extensive than within their existing or prospective blocs. Interbloc trade liberalization would probably be concentrated on those industrial products on which tariffs are already low—say, under 20 percent. It is less probable that products with high degrees of protection would be substantially liberalized because even small cuts in their rates would lead to disproportionately large increases in imports, stimulating strong protectionist pressures from business firms and workers adversely affected. Indeed, such reactions will undoubtedly compel one or another bloc from time to time to reimpose tariffs or quotas on some products previously liberalized. Whether presently protected agricultural com-

modities would be included in interbloc liberalization would depend
upon whether the long-term demographic and economic changes in
Western Europe permitted increasing imports from North America.
With respect to nontariff barriers and trade distortions, the work already underway in the OECD and the GATT would result sooner or
later in reasonable harmonization of the most important of them, that
is, government procurement policies, product standards and testing
methods, business practices and regulation, export-promotion devices,
and environmental quality controls.

Thus, through negotiated tariff reductions and codes of behavior
governing commercial practices and nontariff distortions of trade, the
conditions of *micro* competition among the three major groupings of
the Atlantic region would be increasingly equalized. True, Japan lags
considerably behind the other two in its liberalization of trade and
investment. But, its growing economic strength, self-confidence and
urge to play a more important role in world affairs, combined with
pressure from the United States and the Community, are likely to impel
it to lower these barriers at an accelerated rate in the next few years.

Progress in dealing with the *macro*-adjustment problem would also
have to be within the limitations imposed by the new nationalism. The
gradual emergence of large economic blocs within the region is not
only being fostered by this problem but should also help to ease the
difficulty of coping with it. Bloc formation would do so, moreover, by
means which, unlike the establishment of a regional central bank,
would not require supranational authority. Nonetheless, because of the
ambivalent attitudes and interests on both sides of the Atlantic, these
developments in monetary relationships are likely to come about by
minimum steps and only under the pressure of necessity.

Bloc formation would facilitate greater flexibility of exchange rates
between the dollar, on the one hand, and European currencies and the
Japanese yen, on the other, thereby further improving the functioning
of the international monetary system and making the *macro*-adjustment
process less onerous. Expanded creation of SDRs and the conversion
of all or some of the excessive reserve-dollar accumulations into SDRs
would ease the liquidity problem and help to sustain confidence in the
system. However, even in combination, flexibility of exchange rates
and increased liquidity would be insufficient to cope with the *macro*-
adjustment problem if substantial dollar deficits persist. In that case,
reliance would also have to be placed from time to time upon more
or less restrictive controls over speculative flows of funds among the
blocs. Such measures would be coordinated within each bloc and

might be subject to negotiation among the blocs. They might agree upon means for restoring and maintaining a limited convertibility of the dollar into gold—as well as into SDRs—in official interbloc transactions. The blocs might also cooperate in regulating the size of the Eurocurrency market and the rate and kind of credit creation in it, which would permit more effective control of the volume and volatility of the funds moving into and out of the dollar, especially short-term capital.

In these ways, the amount of dollars in monetary reserves could be more nearly equated to the growth of demand for them, and the anxiety-provoking characteristics of imbalances would be significantly decreased, with consequent diminishing of the tendency toward monetary crises of confidence. The international role of the dollar would be further reduced if restrictions on interbloc dollar flows had to be imposed frequently and if the exchange rate of the dollar fluctuated widely. Such developments would foster the growth and diversification of money markets within the European and Japanese blocs and the use of some of their currencies (in addition to sterling) for reserve and other purposes. Although it is possible that, despite these means for easing *macro* adjustments, excessively large American balance-of-payments deficits could provoke a critical situation, the tripolar organization of the Atlantic economic system would lessen the risk of a world money panic and subsequent depression. For, unlike the situation in 1971, it would be possible for the United States to adopt temporary interbloc trade and exchange controls without undue disruption of economic relations within the blocs.

These developments in monetary relationships among the blocs could also foster the cosmopolitanization of multinational enterprises. On the one hand, restrictions on interbloc dollar flows and uncertainty regarding the exchange rate of the dollar could inhibit the repatriation of earnings and capital by American multinational corporations, especially if returns on investment were higher abroad than in the United States. On the other hand, the European and Japanese governments could decide to reduce, or limit the growth of, their accumulations of reserve dollars by purchasing themselves, or permitting their citizens to buy, increasingly large amounts of the equity securities of U.S. multinational companies, thereby helping to internationalize the latter's ownership. In turn, these trends would intensify the pressures on U.S. corporations—and by example on European and Japanese firms as well—to promote people of more diverse nationalities to top managerial positions. Thus, the two respects in which the internationalization of these enterprises has hitherto been very slow would accelerate during the coming decades.

The emergence of economic blocs and their interactions along the foregoing lines could certainly preserve a very substantial and mutually beneficial degree of economic integration among them. But, such an outcome is not assured. The possible danger arises from the fact that, although trade among them would continue to grow in absolute terms, each large bloc would tend to become more self-sufficient in the aggregate as trade within it increased much faster. Hence, interbloc trade would be of declining *relative* importance. Whether this trend toward self-sufficiency would make it easier or more difficult to settle economic conflicts among the blocs would depend upon the prevailing political and psychological environment. In the 1950s and early 1960s, common interests in European and Japanese recovery, the common external menace of the Soviet Union, and the strong U.S. sense of mission to provide leadership toward a new and better world political and economic system predisposed the United States and its allies to avoid serious quarrels and to make mutual concessions over economic differences. In contrast, the tendency since the late 1960s has been toward more self-concerned and self-protective attitudes in the United States, the European Community and Japan as the common external menace has appeared to recede and as each has become more intently focused on its own internal problems and transformations.

One possible development of this trend could be that protectionist pressures in the United States would become so strengthened by widespread frustration of unrealistic expectations regarding European and Japanese behavior as to compel extensive reimposition of tariffs and quotas. Or, because American policy makers continued to seek a different type of world and regional economic integration—one based not on differential degrees of discrimination but on universal application of the most-favored-nation principle—the United States might reject or delay the kinds of trade, monetary and other cooperative measures needed to sustain economic integration among emerging blocs. Equally possible, these measures could be resisted by the European Community owing to its inability to continue the process of European integration, let alone that of unification. For, the Europeans would be unwilling to participate in interbloc economic integration without a growing sense of their own cohesiveness, identity and strength.

An alternative projection, therefore, would envisage policy makers and opinion leaders on both sides of the Atlantic becoming less and less willing or able to undertake the kinds of actions required to maintain regional economic integration. Instead, the increasing severity of *micro*-competition and *macro*-adjustment problems, magnified by the internal pressures of the new nationalism, could generate more divisive trade disputes and disruptive monetary crises. These reactions could have even more serious consequences, because implicit in the process of bloc formation are conflicting designs of world political and economic order, as well as strong temptations to compete for the affiliation of Asian, African and Latin American countries. Sooner or later, mounting anxieties, resentments and controversies would lead to a vicious spiral of mutual restrictions and retaliations among the blocs, ending with their regression into neomercantilist competition analogous to that prevalent during the great depression of the 1930s.

However, this alternative has a lower probability than the first projection, namely, that the blocs will be able to cope with their *micro* and *macro* difficulties in ways that at least preserve and possibly improve economic integration among them. For, the latter outcome is fostered both by the self-reinforcing characteristic of economic integration and by the more organic nature of the ties holding the blocs together in the various institutional and cultural manifestations of the transnational integration of production. Thus, on economic grounds alone, its chances are considerably better than even. Nevertheless, because economic factors are rarely, if ever, determinative *per se,* the future of the Atlantic region will in the last analysis be shaped within the broader framework of transatlantic relationships in all of their relevant dimensions.

The Prospects for Atlantic Relationships

If the trend in Atlantic economic relationships has been toward the narrowing of disparities, that in political-military relationships has so far changed very little. The disparities in military power and hence in world political influence between the United States and Western Europe have not yet been reduced substantially. And, just as the lessening of economic disparities plays a major role in shaping the future of the regional economic system, so, whether or not political-military disparities persist will help importantly to determine the future of transatlantic relationships in those respects.

The reduction of the one and the continuation of the other type of disparity reflect Western Europe's progress in economic integration and inability to move ahead in political unification. Its own high degree of economic integration makes possible the Community's participation in Atlantic economic integration and, conversely, deters it from joining in Atlantic economic unification. Similarly, the existing extent of Atlantic political-military integration, as represented by NATO and the U.S. nuclear guarantee, has been both a prerequisite for European economic integration and a hindrance to European political-military unification. Thus, European and Atlantic integration and unification are at the same time mutually supporting and mutually incompatible. The Atlantic region can either resolve or live with this basic contradiction in three possible ways.

The first possibility is that the will to achieve superpower status would become so strong among European elite groups and the effects on resource allocations sufficiently acceptable to the people generally to overcome the major impediments to economic and political union. Capable of and bent on pursuing independent interests and objectives, a federal Europe would undoubtedly regard a subordinate status in NATO as no longer necessary for its safety and a hindrance to realization of its policies. Indeed, the likelihood is that a continuation of an integrated alliance even on the basis of equality would be neither possible nor desired. Federal Europe, as the more sensitive and impatient newcomer to superpower status, would undoubtedly resent the mutual constraints of such an arrangement. As the old saying goes, "he who has a partner has a master."

In any case, two superpowers more or less comparable in economic and military strength and in their determination to play active and important roles in world affairs are likely to participate in an integrated alliance only if they are both intent upon achieving a common goal, positive or defensive, of overriding significance. Other interests either do not require or would be frustrated by so close a political-military relationship. The two Atlantic superpowers would find increasingly irksome the formal obligation to refrain from working against each other in situations where each was seeking to promote a competing design for world political and economic order, pursuing conflicting political or economic objectives in Asia, Africa or Latin America, and— above all—insisting on freedom of action to protect or advance its own interests *vis-à-vis* the other superpowers, the Soviet Union, China and Japan. Hence, if Europe united and became a superpower, NATO would very likely be terminated by mutual agreement or reduced to a conventional nonaggression pact. This is why the Grand Design of Atlantic partnership would undoubtedly be outside the limits of the possible, resting as it does on the rationalistic assumption that acceptable compromises could always be reached between the United States and a united Europe because their interests and objectives would "naturally" be the same.

The second possibility is that Europe would not unite but that one or more of the larger countries would separately achieve proto-superpower status. In that event, the fate of the alliance would largely be determined by the nature of the sense of mission of the new proto-superpower and by the state of relationships with the Soviet Union. The new proto-superpower might renounce membership in NATO for essentially the same objections to its constraints as would a federal Europe. France was beginning to do so under de Gaulle. Or, if Germany were to become a proto-superpower, its relations with the Soviet Union, as well as those of the United States, could be sufficiently antagonistic, as explained in Chapter IV, to maintain a close American-German tie and assure preservation of NATO. Even though the other West European members might be increasingly unhappy over growing German power and influence, they would be precluded from seceding from the alliance by their continuing greater fear of the Soviet Union. And, if Germany were to leave NATO to become an independent proto-superpower, such an action would alarm the other European members sufficiently to make them cling to the U.S. guarantee. Hence, even with the emergence of a separate national proto-superpower in Western Europe, the probability is that NATO would persist indefinitely, although with some changes in membership and functions.

If neither a federal Europe develops nor a separate European proto-superpower emerges, the remaining possibility is that European economic integration and political-military coordination would continue to be intensified but by means that would not involve significant political unification. And, the analysis in Chapter V indicates that this essentially confederal possibility has the highest probability. Under it, major changes would be likely to occur in NATO.

The reason is that attitudes and interests on opposite sides of the Atlantic would continue to diverge significantly even in the confederal projection. The decline of both the direct Soviet menace to Western Europe and the threat that recurrent political crises in other parts of the world would trigger a nuclear war between the superpowers has already raised serious questions in Canada and the European members of NATO about the need for the existing degree of military integration and political coordination with the United States. They tend more and more to see the likely external dangers that necessitate continuation of NATO as being threats only to the Atlantic region in the strict geographical sense of the term. Their views are in contrast to the global American conception of the scope of Atlantic security and political interests and hence of the obligations of its NATO allies to support its activities elsewhere in the world. Moreover, although insufficient to overcome the obstacles to European federation, the increasing technocratic positivism and slowly reviving sense of mission of European elites, the self-confidence engendered by the progress of European integration, and the diverging interests and objectives inherent in the process of economic-bloc formation would combine to strengthen the already evident tendency of Europeans to repudiate American leadership and influence, and more and more to insist on determining their own goals and priorities, externally as well as internally.

The changes in NATO would probably be initiated by the growing recognition on both sides of the Atlantic that the rationale of graduated deterrence has very little relevance to the kind of conflict situation which would be likely to arise in Europe and threaten to trigger a nuclear war.[10] One consequence would be that U.S. troops stationed in Western Europe would be cut drastically, the number left depending on how many were psychologically necessary to convince both the Russians and the Europeans of the credibility of the U.S. guarantee. The need to maintain some of the military bases and capabilities no longer supported by the United States and the desire for greater independence of American leadership would then induce the Europeans to make larger defense efforts of their own, thereby fostering political-military coordination (as projected in Chapter V). In the shorter term, these developments would result in the establishment of a second decision center in NATO that would be the political-military parallel of the economic decision center already emerging in consequence of the process of bloc formation. Although inferior to the military power of the United States, it would significantly limit American freedom to determine alliance policy. But, as the second-strike nuclear capability of confederal Europe developed, the Europeans would no longer be satisfied with a subordinate status. Over the longer term, therefore, NATO would likely be converted into a traditional type of mutual assistance pact, with little—perhaps largely nominal—integration of armed forces even at the command level and without the stationing of U.S. troops and bases in Europe.

It remains now to assess the relative probabilities of the different ways in which the transatlantic relationship could develop in all of its relevant dimensions. The possibilities can be conceived as within a range whose limits are, at one extreme, an Atlantic economic and political union and, at the other, a disintegration of the regional system into increasingly antagonistic American and European superpowers.

The Atlantic union end of the range has a quite low probability. It is difficult to imagine developments that would so impair Americans' sense of the adequacy of their nation-state as to make the United States willing to join an Atlantic union, in which its influence would not be paramount and its freedom of action would be substantially restricted. Indeed, even an arrangement by which the European countries would accede to the existing federal union of the United States would probably be bitterly resisted by large sections of American elite-group and public opinion, and might not be approved by the Congress. Conversely, any kind of Atlantic federation in which the United States would be predominant would be regarded by most Europeans as an American empire, not as a union of equals, and they would be unlikely to join it voluntarily. Only an external menace of such magnitude and imminence as to make the alternatives of uniting or perishing the unequivocal choice could overcome these serious obstacles to an Atlantic union. The latter development is among the least probable in the foreseeable future.

The other possible extreme of the range—that transatlantic relationships would deteriorate into superpower rivalries and neomercantilist conflicts between a federal Europe and the United States—has almost as low a probability. Even if Europe were to unite and become a superpower, the likelihood would still not be high that its relationships with the United States would be as universally antagonistic and suspicious as were American-Soviet relations during the cold-war period. The existence of other rival superpowers—the Soviet Union and China—would alone preclude that kind of single-minded bipolar confrontation. Nor would the economic and political issues, however serious, likely to arise between a federal Europe and the United States generate the same degree of substantive irreconcilability and compulsive distrust as those that persist in American-Soviet relations. The basic sociocultural affinities of the Western nations and the differences between them and the other societies and cultures on the planet would help to inhibit such an outcome.

Much the most probable development, therefore, is that transatlantic relationships would continue in the middle of the possible range. If Europe were to federate, relations would very likely tend toward the divisive end. If European integration evolved into a confederal form, they would probably remain on the integrative side, where they are now. But, even in this case, transatlantic relationships are unlikely to be nearly as close, harmonious or easy to manage as they are assumed to be in the expectations of American rationalistic and sentimental parochialisms.

Unless a common external menace were to arise, resolution of the economic and political issues recurrently generated by the basic disparities and inconsistencies in the transatlantic relationship would be strongly affected by psychopolitical factors. The less unified or integrated Europe is, the more the remaining disparities between it and the United States will weaken its self-confidence and heighten its frustration and resentment. But, the more unified or integrated Europe is, the more its increasing self-confidence and sense of mission will reduce its willingness to make concessions to the United States. This reaction will be especially marked in the upcoming generation of elites who will reach top-level policy-making and opinion-molding positions in the late 1970s and '80s. They will be more thoroughly imbued with technocratic positivism and the sense of vocational mission than the present generation of European leaders. And, unlike the latter, their attitudes will not have been indelibly colored with self-doubts by the experience of national inadequacy and extreme dependence on the United States in the postwar period of cold war. On the other side of the transatlantic relationship, the disappointment of American anticipations of cooperative European behavior and the relative decline of U.S. power and influence in the international system will lead to growing unwillingness on the part of the United States to meet European expectations of American concessions—to their weakness if unification or integration lags and to their strength if either progresses. Thus, the propensity on both sides to settle conflicts by compromise will be diminished. It is impossible to predict the specific events in which transatlantic issues and disputes will manifest themselves in the years to come. But, their general characteristics, seriousness and frequency can be forecast, as in this chapter, because their sources are inherent in the nature of the current period of world order and of the development of Western society.

In the foreseeable future, therefore, the Atlantic region will be troubled by more perplexing questions and less tractable problems than those it has had to endure in the decades since World War II. And, on both sides of the Atlantic, these difficulties will be all the more exasperating to policy makers and opinion leaders increasingly motivated by positivistic convictions of their mastery over nature and society. Serious as the resulting conflicts will undoubtedly be, however, they are not likely to be so severe and prolonged as to engender a fundamental antagonism and basic distrust. Such an outcome is inhibited by the pervasive cohesion of sociocultural affinities, the organic nature of the economic ties expressed in the transnational integration of production, and the persisting common interest in each other's economic well-being and political freedom. Thus, although economic disputes, political controversies, and psychological anxieties and frustrations will plague the transatlantic relationship, they will not fatally undermine the security, prosperity and dynamism of Western societies.

Both of the parallel processes of European and interbloc integration —in part competitive and in part mutually supportive—would involve increasing restrictions on the freedom of action of national governments. Such a trend would mean that, along with the strengthening of the domestic institutional bases of Atlantic nation-states, their scope for conducting independent policies and actions would be narrowing. In effect, they would be exercising their sovereign powers more and more in common. This development differs sufficiently in its organizational and operational manifestations from the deliberate transfer of crucial functions to supranational authorities for it to be much more acceptable politically and psychologically in the period of the new nationalism. Hence, while the institutional roots of nation-states would continue to spread wider and deeper within their own societies, their branches would grow more and more intertwined and interdependent.

If these trends should continue for the remainder of the century, it could well be that new forms of *macro*-social organization would imperceptibly evolve that, in accordance with the paradoxical nature of human history, would in quite unintended and unexpected ways both preserve the diversity and strengthen the unity of the countries involved. Such possible *macro*-social changes would mark the end of the current period of the new nationalism, at least for the Atlantic region. Whether and in what circumstances the independent sovereign nation-state might eventually pass away are questions whose determinants within Western societies and in the regional and worldwide systems lie beyond the range for which an empirically based projection can be made. Some speculations on how such developments might occur are offered in the next chapter. In any event, it is probable that, for all their continuing tensions and problems—indeed, in part because of them—Western society and culture will still be capable of great creative acts of innovation and statesmanship when the times are again propitious for them.

VII

Glimpses into a Possible
Twenty-First Century

ALL OF THE PROJECTIONS presented in preceding chapters deal with the more and the less probable courses of development at national, regional and worldwide levels during the remainder of the recently begun period of the new nationalism. Having looked ahead for several decades, we turn to the next question: What lies beyond the new nationalism for the Atlantic countries? How are the determinative characteristics of the present period likely to be so transformed or replaced in the coming years as to bring it to an end? What, if anything, can be foreseen of the fate of Western societies in the 21st century?

An attempt to look so far forward encounters difficulties more formidable than those that hampered the projective efforts in preceding chapters. Those projections are in the main derived from analysis and interpretation of established determinative trends and of the already evident influences that are likely to modify them significantly in the next two or three decades. Thus, to the extent possible in forecasting, they are based on empirical data. In contrast, a projection of developments beyond those of the present period can have no such basis. By definition, it attempts to analyze the characteristics of a time whose determinative trends and modifying influences can at best be discerned only in vaguest outline. Moreover, the longer the time horizon of a projection, the greater is the scope for unforeseeable novelty.

In consequence, the analysis in this chapter is much more speculative and formalistic than that in previous chapters. The method used is to extrapolate briefly the two extreme limits of the possible in the next period by logical deduction from alternative sets of trends, both manifest and barely incipient in the present period, which could become of determinative importance by the end of the century. Then, an effort is made to work out in greater detail a more probable consistent set of possible developments in the middle of the range between these extremes. For this type of projection, the only supporting empirical verification would be the evidence of how similar societies have behaved in more or less analogous circumstances in the past—a test that is partly a matter of interpretation.[1]

Even more than in Chapter II, the focus here is on the future possibilities of American society and culture. The reason is that, just as the United States led the other Atlantic nations in starting upon the technocratic phase of Western civilization, so will it precede them in the further development and eventual transformation of the characteristics of the current period. It is reasonable to suppose that, because of their substantial sociocultural affinities and similar directions of evolution, the other Western societies in the Atlantic region would sooner or later experience changes similar to those projected here for the United States, even though they might differ in their specific forms, emphases and timing.

The Extremes of the Possible: The Decline and Fall

The most unfavorable projection is based on the assumption that the worst possible fears for the future of American society would be realized. The socially disintegrative manifestations that so alarm people today would not only continue during the remaining decades of the century but would gradually become the dominant trends. This would mean that, by the opening years of the 21st century, the technocratic society of the United States would be in a more or less advanced stage of disintegration. In such circumstances, the contemporary prophets of doom, who are in good part motivated by the desire to spur actions to prevent the fate they predict, would turn out to be much more accurate than they wish.

The Process of Decline

The main characteristics of such a long-term process of social disintegration can readily be extrapolated from some of the current phenomena and trends analyzed in Chapter II.

A worst-outcome projection would have to assume that little, if any, progress would be made in eliminating relative poverty and in integrating racial and ethnic minorities and other disadvantaged groups more fully into American society through better education, adequate employment opportunities, and higher living standards. Moreover, the numbers of unemployed within those groups would tend to increase owing to the rising level of skills needed for work in industry and even in the service trades, and to labor-saving technological innovations. These depressed groups would be regarded with indifference by, and their efforts to improve their condition would meet with hostility from, the regularly employed blue- and white-collar workers, on whom the economic burden of remedial measures would fall most heavily through taxation and inflation. Prolonged frustration of aspirations for a better life would lead the members of the depressed groups increasingly to socially disintegrative responses. Sporadic mass rioting, looting and arson; individual crimes against persons and property; recruitment into criminal syndicates and neighborhood gangs; organized revolutionary activity; drug-taking, schizophrenic flights from reality, and apathy would become more and more widespread. And, so, too, would repressive measures for dealing with them.

Concurrently, the numbers of youthful dissenters and drop-outs from the elites and other adequately employed groups would tend to grow. In the early stages, economic security, greater parental indulgence and permissiveness, the increasing ineffectualness of the institutions for popular education, and the long postponement among the elite groups of self-responsible adulthood until the completion of professional training would more and more weaken the effectiveness of internalized behavioral norms and external social restraints and strengthen the pressure for immediate gratification of desires and objectives. Such satisfactions would increasingly be sought in direct, sense-saturating and highly dramatic forms owing to the influence of prolonged exposure during childhood and adolescence to the audio-visual media and other kinds of popular entertainment. These vivid, isolating and time-consuming experiences would reduce the opportunities for socialized play, in which much of the capacity for forming intimate and satisfying interpersonal relations is initially developed. Hence, adolescents and young adults would seek, but usually would be unable to sustain, warm peer-group relationships, and they would be impelled to react to the frustration of their desires in emotional rather than rational forms. Resolution of conflicted feelings toward parents and other adults would be especially difficult for elite-group young people. On the one hand, parents and teachers would still endeavor to inculcate in them the values of social responsibility, personal integrity and pursuit of knowledge and progress. On the other hand, elitist adolescents would witness the denial of these values in the dehumanization and materialism of the technocratic society and in the deceptions,

fears and frustrations inherent in its competition for personal achievement and occupational advancement.

As time went on, these and other difficulties of the maturation
process would be effectively resolved by fewer and fewer children, and
even they would conform to the life-styles of their parents with a
lessening sense of redemptive mission and vocational conscientiousness.
Conversely, a growing number would be unable to do so, experiencing
feelings of inadequate identity, a sense of meaninglessness, of alienation from self and society. These reactions would lead them to seek
various forms of isolated, individualistic or philadelphic ways of living,
or would be expressed in violence, withdrawal and other kinds of
socially disintegrative behavior.

The effects of these trends would be aggravated by the continued
deterioration of the physical environment, with its adverse impact on
health and morale. Urban rehabilitation, control of the many forms of
pollution, and provision of public services and facilities would be less
and less adequate relative to needs, especially for the lower-income and
depressed groups. Slums would grow faster than they were replaced
and the cities would continue to decay.

These difficulties would gradually intensify social conflicts between
the elites and the regularly employed portions of the population, on the
one hand, and the depressed groups and dissenters, on the other. However, these developments would also tend to increase disputes among
the elites and between them and the regularly employed. As social
problems worsened, contentions of the latter type would grow concerning the kind and cost of measures for dealing with the mounting
difficulties of the society. These conflicts would, in turn, further
strengthen the disintegrative trends.

One consequence would be the increasing role of violence within the
society. It would become more pronounced among the members of the
depressed and dissenting groups, individually and in the collective
forms noted above, as well as on the part of the law-enforcement
agencies. Also, other government departments and the major private
institutions and organizations would more and more be impelled to
resort to violence, first in self-defense and then, as they became habituated to the use of force, as a means for advancing their own interests.
Official agencies, political parties, large corporations, trade unions,
residential protection organizations, even universities and religious
sects and congregations, would gradually develop their own security
forces and fortified premises. Neither city streets nor suburban roads
would be safe at night, and many not even in the daytime, as crime

increased, mass riots became more frequent, and armed conflicts between groups and organizations began to spread.

The increasing disorder of the society would sooner or later adversely affect economic growth, disrupting economic relationships and making markets more uncertain. In consequence, business corporations large enough to command the necessary resources would be impelled to assure their access to the factors of production and to markets by mergers, hegemonies over smaller companies, alliances with banks and trade unions, and pressures on the government. Competition among them would become more intense as they sought to protect and advance their interests by both peaceful and forcible means. And, as they succeeded in their efforts to control more and more of the needed production factors and markets, their power *vis-à-vis* that of the government would increase. In this projection, these trends would exaggerate the cosmopolitanization of multinational corporations, making them into quasi-sovereign "principalities" increasingly independent of national governments. Similar effects would be evident in the behavior of trade unions and other private and governmental organizations.

The growing predominance of these trends would mean that the pluralism of 20th-century American society would disintegrate into particularism in the 21st century. This would be a reversal of the historical process of development characteristic of Western society since the early medieval period.[2] The influence of universalistic values and of the sense of the public interest, of the common good, in integrating diverse groups and institutions and orienting them toward generally accepted goals would decline. And, as in particularistic societies, loyalties to closer and more concrete sources of power and assistance would tend to displace those to more distant and generalized authorities, such as the central government and the nation-state. The increasing need for physical protection, the fear of losing economic status and the impossibility of maintaining or improving it in the absence of educational and occupational opportunities, and the weakening of the individual's sense of identity and security would impel him to seek or accept permanent attachments to the particular institutions and organizations capable of meeting these needs. In return, the latter would expect unquestioning loyalty and obedience, which would strengthen the commitment to them of their members, employees and client organizations.[3] Whether personalism, too, would revive with particularism would depend upon whether the force of a prepotent leader's personality would be necessary to make technocratic institutions effective in their new patronal role.

As they became the predominant characteristics of the society around the turn of the century, these trends would in essence bring about a kind of *new feudalism*. It would not be comprised, like historical feudalisms, of a loose hierarchy of authority-subordination relationships based on autarkic landed estates. But, it would resemble them in the relative decentralization of power in more or less autonomous social units, commanding the primary loyalty of their members and providing them with the protection and the resources and skills needed to preserve their personal security and economic status. The emergence of the new feudalism would mean a reversal of the long-term trend of increasing central-government power characteristic of the present period of the new nationalism. However, the functions now carried on by central governments would not cease to be performed. As explained above, some would be exercised by other institutions and organizations. Others would continue to be discharged by government agencies but in an increasingly independent and self-interested manner. That is, the activities of the departments of defense, of internal security, of welfare services for the depressed groups, of environmental control, and others, would be less and less coordinated and effectively supervised by the top executive authority. The distinction between public and private would tend to become meaningless, as it was in historical feudal regimes. Private organizations would acquire quasi-official powers, and public agencies would more and more behave like them in particularistic ways—planning, maneuvering and struggling to protect and advance their institutional interests and those of their dependents. Thus, the new feudalism would not mean the disappearance or atrophy of the central government, as feudalism did in the past.

The new feudalism would both reflect and reinforce the gradual stagnation of scientific and technological development. Staffed by technocrats whose sense of mission and occupational conscientiousness would be declining, the powerful semi-independent organizations would be less and less impelled to innovate and would be constrained by deteriorating conditions to use their resources for more immediate purposes. Basic scientific work would especially suffer as the economic resources and educational facilities needed for it became scarcer and research was more and more limited to protecting or advancing the interests of the institutions providing funds for it.

In turn, declining scientific and technical research would sooner or later result in a falling rate of productivity growth, which would further reduce the resources allocated to developmental purposes. Reinforced by the disintegrative trends already described, this vicious spiral would also be accelerated by the effects of inadequate technological innovation on raw-material costs. For, as the high-yield sources of materials were exhausted, the rising real costs of producing or importing them from low-yield sources would get so far out of line with the stagnating real national income that the latter could no longer be maintained and would thereafter steadily fall. This economic deterioration would hasten the process of social disintegration.

Nor would the advent of the new feudalism be prevented by a period of totalitarian rule. Indeed, if the society survived a totalitarian regime, the latter's inevitable failures and after-effects would strengthen the trends leading toward the new feudalism. In the United States, Great Britain, Canada, the Netherlands, Switzerland and Scandinavia, efforts to impose totalitarian rule would probably be inhibited by their democratic traditions for long enough for such regimes, when they finally came to power, to be unable even briefly to check social disintegration. Or if, as in central and southern Europe, they were established earlier in the disintegrative process, dictatorships could at most temporarily arrest but not reverse the decline of the society. A totalitarian regime would be responsive to the interests of the semi-independent and armed organizations of the elites and other established groups and would have to rely upon them in substantial degree to cope with disintegrative problems and to carry out many of the measures of social control. Hence, even if 21st-century totalitarian regimes were to be long-lasting, as they might be in most of the continental countries, with their strong traditions of superordinate state power, they would tend to be less unanimist and centralized than those of the Soviet Union and the East European nations today. Their significance would be analogous to the revival of the centralized autocracy by the fourth-century Roman emperors, whose efforts temporarily halted the decline but whose methods eventually contributed to its resumption.

The Fall

So much for the bare outlines of a possible process of decline. How might it eventuate in a fall? There would be two possibilities: war or internal collapse, the first more likely than the second. The reason is that, despite mass discontent and violence, feudalistic regimes and semi-independent business organizations would be motivated by their interests to cooperate in enforcing the minimum degree of order and calculability required to maintain production at a level that could provide at least barely tolerable living standards for the depressed groups and substantially higher incomes for the elites and regularly employed portions of the population. Conversely, the greatest likelihood of internal collapse would arise from the failure to prevent violence among groups and organizations from escalating into a nationwide civil war that could end in general disintegration.

The more probable outcome—war with other countries—might occur in several ways. During the decades in which the new feudalism was developing and even after its predominance, the direct and indirect interests of the semi-autonomous institutions and groups would undoubtedly combine to preserve a sufficient nuclear capability to prevent a Soviet or Chinese attack. A major war would be more likely to arise, therefore, from actions by the United States. The regime, especially if totalitarian, might seek to reunify the society and divert attention from domestic problems by intensifying the sense of an external menace or by aggressive efforts to achieve world- or region-transforming goals. The Soviet Union, China or one or more proto-superpowers might feel sufficiently threatened or thwarted by these activities to enter into direct confrontation with the United States, leading to nuclear war. Another possibility would be that a prolonged American civil war would tempt a superpower or proto-superpower to intervene, or to seek to conquer the United States—actions that could also result in a nuclear war.

Why Unlikely

These ways in which the eventual collapse or destruction of American society might come about have a fairly low probability. I would also assess as small the likelihood of the entire process of decay leading to the fall; indeed, as barely within the limits of the possible. There are several reasons for these conclusions.

Essentially, they relate to the existing characteristics and prospective development of the technocratic society. Redemptive activism, positivistic faith in the power of reason and science, commitment to the use of rational calculation and efficiency criteria, and expectations of continued progress toward eventual social perfection and individual happiness are so strongly imbedded in the institutions, values and processes of socialization and acculturation of Western societies as to make these modes of thinking and acting increasingly predominant in the decades to come. Despite resentment and indifference, the growing effectiveness of humanistic values will continue to strengthen the sense of social responsibility and the conviction that all people should have equal opportunities for the pursuit of happiness. In consequence, at least minimum success in coping with existing and emerging problems is much more probable than the failure depicted in the foregoing pages.

In Chapter II, it was pointed out that most of the dissension in contemporary Western societies is not over the nature of the technocratic order *per se* but over specific national and group goals and over the effectiveness and moral validity of the means for achieving them. During the coming decades, the growing availability of resources for realizing these objectives will help to prevent conflicts from becoming so severe as to lead to the irreversibly disintegrative trends described above. Also contributing to the avoidance of such an outcome is the fact that knowledge of how to deal with present and prospective problems is likely to improve substantially. American pragmatism sooner or later leads to better understanding of the complexities and perplexities of reality, initially ignored under the influence of simplistic rationalism and overconfidence in technical prescriptions. Hence, during the remainder of the century, significant progress is likely to be made in mitigating, if not completely removing, such problems as poverty, environmental pollution, urban decay and educational deficiencies.

It is also likely that the motivations and capabilities of the majority of blacks and other deprived groups will be so changed in the course of the next two or three decades that they will share more equitably in the fruits of an increasingly affluent society. This implies that the current search for a separate identity on the part of militant black leaders is in all probability doomed to failure in its own terms, although it fulfills another essential psychosocial function, that of protecting against inadequate self-respect and self-confidence. The elements of a separate cultural tradition still retained by American negroes are insufficient for constructing a new distinctive identity. And, the attractions and prestige of the prevailing white culture are irresistible, not least because the blacks' own socialization and acculturation processes have for too long been too deeply affected by those of the white society within which they have been living for so many generations. Hence, most blacks aspire to, and increasing numbers of them are achieving, a life-style like that of white "middle" Americans.

Radical dissent from the technocratic society will probably continue to be an important characteristic of the coming decades, and its future evolution and significance will be discussed in a later section. However, it is not likely to be of such nature as to lead to the decline and fall described above. Moreover, it should be recognized that most radically

dissenting young people sooner or later accede to the dominant society
and culture, voluntarily or under the pressure of family and other responsibilities. As with the radical generation of the 1930s, their youthful experiences will predispose many of them to support movements
for social reform, although theirs will be different from those of their
parents. They are also likely to be more tolerant of deviant life-styles
and more relaxed about the dissenting activities of their own children
than are their parents, the radicals and liberals of the 1930s, in whom
the insecurities of the great depression and the influence of puritanical
Marxism and socialism still linger.

The Extremes of the Possible: Technological and Humanistic Utopias

In the strict meaning of the term, utopias are not, of course, within
the limits of the possible: by definition, they are perfectly ordered
societies without serious social or personal problems. However, they
play a key role in one of the longest historical traditions of Western
civilization, serving not only as ideal models with which to contrast
existing social deficiencies but also as program goals to be achieved in
the here and now. Both aspects continue to be important today, the first
as a guide to the outcome confidently anticipated by those who believe
that the progress of science and technology and the power of reason
guarantee social perfectibility, and the second by those seeking to realize humanistic values by living in small philadelphic communities.
These expectations justify broadening the meaning of the term to
designate the other extreme of social development—the possibility
envisaged by the optimists that, in one way or another, all of the
deficiencies of society will be corrected in the future.

The technological and humanistic visions of continued social progress differ radically in their basic principle of organization, although
both anticipate that man will soon be truly the master of his fate. In
the technological utopia, scientific and technical advances and the impersonal application of efficiency criteria in decision making will assure the ability to cope successfully with all present and prospective
difficulties and will make possible continuously rising living standards
for healthy, happy, rational people in a peaceful and well-ordered
society. In the humanistic utopia, the power of moral and aesthetic
values to guide human actions will be so enhanced as to enable people
to live together in harmony with little, if any, coercive authority or
social repression, while each individual is free to realize his unique

potentialities for self-development. The one emphasizes social perfection at the *macro* level; the other, personal fulfillment at the *micro* level. In essence, the first involves high living and plain thinking; the second plain living and high thinking. And, each is in part a reaction to the other in the age-old opposition of reason and emotion, mechanism and spirit, things and ideals.

The Improbability of the Technological Utopia

The nature and rationale of the technological utopia have become increasingly familiar since the early 1960s thanks to the already large and still growing literature forecasting future technological developments and depicting the ways in which their rational application will transform the conditions of life. Hence, there is no need to summarize here the many glowing descriptions of technological progress—the unprecedented advances in electronics, computers and automated production; the faster and more efficient means of transportation and communication; the new and more versatile synthetic materials; the breakthroughs in biology that will make possible genetic engineering and the control of disease, aging and mental illness; and the many marvelous gadgets that are going to make daily living easier, quicker and better. More important, this imminent mastery over the forces of nature is supposed to produce an equal mastery over the forces of society and of the individual psyche. Intent on ridding themselves of the manifold troubles that have hitherto plagued human existence and on enjoying the limitless benefits inherent in technological progress, people will willingly compose their differences over goals and resource allocations, reconcile their competing interests, and hasten the required changes in institutions and behavioral norms—all under the benevolent and disinterested guidance of the technocrats, who alone possess, or will soon develop, the requisite knowledge and techniques. And, those unwilling to act in this rational manner are clearly unable to to so because of mental illness, which will be readily correctable by chemical therapies and personality-changing drugs.

It is highly probable that, in the decades ahead, knowledge and techniques will become available for rebuilding urban housing, transportation and other facilities; for controlling air, water and noise pollution; for improving health, lengthening life, and keeping the aged alert and active; for modernizing and expanding educational systems; for providing recreational amenities needed by growing numbers of people during increased waking time no longer required for work; and for producing many new gadgets that will make living more convenient and pleasant. Also, by raising productivity and improving skills, technological advances will expand the resources for satisfying more of the demands for realizing competing goals, thereby easing some, at least, of the conflicts among them. But, the well-ordered society and well-adjusted people predicted by the technological futurists are not very likely to eventuate for two main reasons.

First, there is always a delay between new discoveries and their application that is not simply a question of the time required for development. True, this period has been progressively shortening as knowledge has expanded and the benefits of faster use have become more highly valued. But, the application of many of the technological advances most significant for the improvements envisaged by the futurists requires substantial resources, and hence is likely to lag considerably behind the completion of research and development. In turn, the effects of these innovations on institutions and relationships will be correspondingly delayed and reduced. It is probable that, even with persisting high rates of economic growth and more adequate manpower training, resources will not be sufficient to apply more than a portion of the available technology to remedying social deficiencies. As demands for resources continually increase in dynamic Western societies, the resulting need to make decisions, explicitly or implicitly, regarding priorities and resource allocations will provide opportunities for those opposed to particular technological innovations and their related sociocultural changes to block, reduce and revise the programs and projects proposed for implementing them.

Second, and much more important, the notion that the progress of science and technology will eliminate all or even many grave social problems and not itself help to produce new difficulties rests on a misconception of the nature of the social process. Innovations in science and technology affect the evolution of societies mainly in two ways: by major increases in the productivity of the economic system and related changes in the relationships among social groups, and by significantly influencing the modes of perception and conception. However, such changes are not the sole determinants of developments in these and the many other areas and dimensions of human experience. Scientific and technological change is in many cases a necessary, but rarely is it a sufficient, condition for social change. The effects of innovations will differ both in degree and in kind, as well as in timing, depending upon the other types of changes that may or may not occur in institutions and values and in modes of perception and conception. Whether people will be willing and able to make the requisite modifications depends on other factors as well as on the promise or the threat of the new technology—the configuration of competing interests involved, the differing prescriptions proposed, the consequences for other important social purposes of diverting resources to make changes, the inertia of institutions, the conservatism of values, and so forth. And, when the innovations do occur, they usually generate new problems either through unforeseen adverse feedback effects on existing conditions or by creating unpredictable new difficulties.

Moreover, although advances in science and technology significantly affect modes of perception and conception, the new ways of seeing and thinking do not necessarily make people more rational, more willing to guide their decisions by impersonal efficiency criteria, better able to prevent egoistic drives and emotional processes from distorting or nullifying the dispassionate weighing of costs and benefits, opportunities and risks. Rather, they tend to influence the dramatic design of the culture, the sense of the society's identity and purpose, and its world view, the dominant perspectives on the nature of the universe and of man's place and function in it. That such changes induced by the progress of science and technology do not inevitably make people think or act more rationally may be demonstrated by the horrors of the 20th century—surely as bad as, if not worse than, those of the 18th or 19th centuries, when the dominant conceptions of society, man and nature were different.

Even if, as in the past, advances in science affect the choice and relative importance of the values to which the society aspires, these innovations do not eliminate the conflicts among such goals. Order cannot be maintained without substantial sacrifice of freedom. Justice cannot be achieved without considerable impairment of brotherhood. Welfare cannot be improved without significant denial of equality. And, so long as resources and opportunities are not unlimited, the fact that people are different and hence have different interests means that there is bound to be competition among them. Answering the 18th-century rationalists, who insisted that political reforms were the sovereign remedy for transforming society, Oliver Goldsmith wrote two centuries ago:

> *How small, of all that human hearts endure,*
> *The part which laws or kings can cause or cure.*

Today, and in the foreseeable future, the same could be said of machines and techniques.

The Unlikelihood of the Humanistic Utopia

Indeed, this is the main reason why humanistic utopias attract adherents. Far from solving the problems of the technocratic society, the progress of science and technology is seen by the humanists as only making them worse. And, they believe these problems to be essentially irremediable in the technocratic society, inherent in its gigantic scale and terrifying complexity, and in its concomitant need for impersonal relationships, efficiency criteria and mechanistic forms of working and living. In consequence, they reject the technocratic society "root and branch" and are willing to pay the price of doing so. Horrified by the dehumanization of mass production and disdaining the materialism of mass consumption, the humanists seek to replace the technocratic order by a society composed of small organizations and communities responsive to the needs and potentialities of the human spirit.

Some designers of humanistic utopias are neoanarchists, especially among the young and their older mentors, envisaging the free expression of human personality in authority-less communities of Edenic innocence and plenty. Most, however, recognize the necessity of formal organization and control. Some envision the constituent social units as largely self-sufficient, reverting to earlier handicraft methods of industrial production and organic agriculture and with commerce among them limited to essential products unobtainable locally. Others foresee no need to renounce the advantages of machinery and electronic equipment, but propose to use only types conducive to small-scale production, in which the worker's skill would be paramount, he would have the satisfaction of making entire products, and interpersonal relationships would be cooperative and not hierarchical.

Even in the latter type of humanistic utopia, productivity would be insufficient to support large populations at reasonably satisfactory levels of living. However, this, too, is regarded by humanistic utopians as a benefit, because a drastic decline in population would relieve the anxieties and hostilities generated by crowding and would make possible the restoration of ravaged natural environments. No longer dominated by dehumanizing work and the compulsion of ever-rising consumption, people would be free to develop their potentialities for self-fulfillment in a wide variety of mutually acceptable ways. These social and psychological changes would encourage warm, loving interpersonal relationships, which in turn would eliminate mental illness and crime. The result would be a well-rounded, diversified society, without serious conflict or unhappiness, dedicated to achieving the nobler possibilities of man as part, not master, of nature.

This picture is undeniably attractive not only to those who reject, but also to many who seek only to reform, the technocratic society. However, it is unlikely to be realized for essentially the same reasons that preclude the social perfection expected by the technological futurists. Neither the complexities of society nor the perplexities of human behavior can be overcome or evaded by small philadelphic communities, as the history of 19th- and 20th-century efforts to establish and maintain them shows. All have sooner or later been dissolved due to these problems, which Fourier and other 19th-century utopian designers strove so hard to eliminate or control. Moreover, it is difficult to imagine that contemporary Western societies, as distinct from groups and individuals within them, would be either willing or able to give up the benefits of high mass consumption, however disturbed many people may become over the other consequences of mass production. A basic revolution in values and attitudes would be required to displace the welfare objective as a major social goal. Such a change normally occurs over generations, not decades.

In sum, neither technological nor humanistic utopias are likely outcomes of the present period. Yet some of the elements of which each is composed will probably play important roles in shaping the characteristics of the new period that will emerge in the 21st century. Those that seem to me most likely to do so are included among the assumptions, explained in the next section, for a median projection between the two extremes.

The Assumptions of a Median Projection

The median projection presented in this and the two following sections is not a prediction of what Western society will probably be like in the new period. It is simply an extrapolation of how a society like ours would develop if certain trends, which are already manifest or incipient, were to become the determinative elements in it by the opening years of the 21st century. The major assumptions regarding these trends are explained in this section.

In reality, however, the trends depicted here are not likely to be the sole important factors at work. I must urge again that, in reading this and the next two sections, the *ceteris paribus* qualification be kept in mind. The social structure and relationships, the motivations and life-styles comprised in the median projection would be more or less modified in real life by many other factors, internal and external. They would include not only the unforeseeable novelties arising in human affairs but also the deliberate decisions and actions of the elites and the people generally. And, it is above all the latter set of unpredictable variables that validates the speculative kind of projection technique used in this chapter. For, if policy makers and opinion leaders can see the logical consequences of different policy choices, they can perceive more clearly the ways in which they could affect the future course of events, and evaluate more accurately the desirability of and prerequisites for the various probable developments.

The first assumption of the median projection is that most of the serious problems plaguing American society today would be substantially mitigated or resolved before the advent of the new period in the 21st century. Difficult as they may be, the tasks of eliminating poverty, integrating disadvantaged groups more effectively into the "mainstream" culture, rebuilding the cities, arresting environmental deterioration, and undertaking the other related reforms are accomplishable with the resources likely to be available within the next two or three decades. Nevertheless, the dangers inherent in such efforts need to be stressed. For example, too rapid or too massive a redistribution of income could have very adverse effects on the U.S. economy, limiting the funds available for investment in the private sector and encouraging large-scale capital flight, which would contribute to industrial and financial disruptions, severe balance-of-payments deficits, and serious unemployment. However, in this projection, it is assumed that the difficulties will be surmounted and the dangers avoided through sufficient realism, moderation and skill in managing the complex economic, social and psychological changes involved. These developments would, in turn, profoundly affect the nature of American society in the 21st century, as explained in the succeeding pages.

The second assumption follows in part from the first. It is that, by the early 21st century, the major divisions within American society would not be into economic and ethnic groups as they are today. The current differentiation would decline in importance owing to the social integration and rising living standards assumed in the paragraph above, and the unlikelihood of new mass migration to the United States. Already in the decades since World War II, the marked ethnic diversities introduced by the large-scale 19th-century immigration, and the different urban and rural life-styles have been gradually coalescing through the dissemination of middle-class standards and behavioral norms by such means as the educational system, the mass media and advertising. Although ethnic differences are still evident and become more important from time to time, the long-term trend is toward their virtual disappearance in the course of the next generation or so. In the process, the Irish, Italian, Slavic, Germanic, Jewish, Negro, Spanish and other cultural elements will continue to be amalgamated with the older Protestant English ways of living and thinking. Thus, it will only be in the late 20th century that the American "melting pot," proclaimed in the late 19th century, finally succeeds in fusing a sociocultural homogeneity equal to that which the United States knew before the great immigration and urbanization that began in the 1840s. The steadily declining significance of ethnic and economic differences by the end of this century would open the way for the new kind of social differentiation described in the next section.

The third assumption is that humanistic criticism of the developing technocratic order will have an important effect on its institutions and behavioral norms by the beginning of the new period. Humanistic utopias would not replace the technocratic society for the reasons already explained. But, humanistic values are integral parts of the Western cultural tradition and their increasing effectiveness in recent centuries has helped to bring about the many social reforms accomplished during the past two hundred years, and especially in the 20th century. The changes likely to be fostered by the growing power of humanistic values in the decades to come include those in the first assumption and in the others explained below.

The fourth assumption, with which technological futurists and many economists would agree, is that forthcoming advances in science and technology and in the capacity to manage the economic system would result in continuing increases in productivity at a high rate. Hence, early in the next century, if not before, the cumulative application of ever more sophisticated computers, computer-controlled machines and tools, and other automated devices would result in a qualitative jump in the productiveness of the economic system analogous to that which took place during the industrial revolution of the 19th century. As this development occurred, the human labor needed in the industrial sector to produce a growing output would decline, and total industrial employment could be prevented from falling drastically only by substantially reducing each worker's time on the job. The growth of the larger and hitherto expanding service sector has depended much more on rising employment than on greater output per worker because the scope for technological innovation is smaller in many service activities. Even so, the possibilities and pressures for labor-saving standardization and computerization, as well as for self-service in retailing and other appropriate consumer fields, would slow down and eventually make unnecessary the growth of total employment in the service sector. This development would lead to reduction of the time on the job of white-collar workers so as to provide additional employment and to match the cuts in the length of the blue-collar working day or week. Moreover, the more standardized, routinized and mechanical the human role in industrial production and the service trades becomes, the more necessary it will be to shorten the duration of the individual's participation in them in order to prevent the conscientiousness of work performance and the quality of output from declining to uneconomic levels.

Unlike in the decline-and-fall projection, the exhaustion of high-yield raw-material sources would not interfere with the production of a growing volume of goods and services. For, continuing scientific and technological research and development would provide a steady stream of new and improved extraction and processing technologies, methods of use economy and waste recycling, and substitutes and synthetics, as well as fusion power. This cumulative technological innovation would prevent the real costs of producing or importing materials from low-yield sources from rising significantly faster than the growth of real national income. Indeed, at successively higher levels of real income, it will become economic first to extract metals and other materials from the oceans, and later—assuming continued advances in space technology—to mine the moon, Mars and even the asteroid belt. If, however, despite research and development, the real costs of producing some indispensable materials were to increase disproportionately to the growth of real national income, the pressure of rising prices would lead to changes in consumption patterns and production strategies. Manufacturers would be forced sooner or later to design consumer products containing substantial amounts of such materials not, as now, for rapid replacement but, as in the past, for durability.

The fifth assumption is intimately related to the preceding one. It is that, by the opening decade of the 21st century, the goods and services required for rising living standards and progressively to meet the other goals noted above (as well as to sustain economic growth within the limits imposed by the need to avoid prolonged, serious internal and external imbalances and ecological damage) would be produced with three or four hours of work per day by—or, less likely, with a longer work day for only 40 or 50 percent of—the nonelite population of working age. Thus the median projection assumes that a substantial portion of the income gains accruing to blue- and white-collar workers would be realized in the form of shorter hours of work, and not solely in the form of increased consumption of goods and services.

Such a preference is already implicit in the trend toward longer vacations, earlier retirements and the beginnings of a move for a three-day weekend, as well as in the rising rates of absenteeism and turnover and the other contemporary evidences of the decline of the gospel of work. True, the additional time has hitherto been used in many cases for "moonlighting"—holding a second job to raise income and living standards. But, in this projection, it is assumed that the shift in con-sumption preferences and the other socioeconomic considerations favoring accelerated automation would be decisively reinforced by the increasing influence of humanistic values—by growing concern for the "quality of life" and revulsion against the mind-deadening routinization of blue- and white-collar work, which loom so large in criticisms of the technocratic society. Thus, by the early decades of the 21st century, the dehumanizing effects of mass production would be substantially re-duced, not by abolishing the technocratic society as the humanistic utopians demand, but by drastically modifying the nature and duration of human participation in economic activities through widespread auto-mation and shortening of working time.

The sixth assumption is that the reduction of the hours of labor will be accompanied by full realization of the current incipient trend in national social-welfare policy toward divorcing incomes from work. Such a development would both reflect and reinforce the more basic shifts in values and attitudes so that, by the end of the century, the moral, socioeconomic and psychological implications of unemployment would no longer be such as to generate strongly negative feelings. In the new period, therefore, a high-priority claim on the greatly increased resources then available would be accorded to guaranteeing all mem-bers of the society at least a minimum income sufficient to assure them an adequate standard of life, which would be considerably above today's level. European countries are already moving in this direction and the United States and Canada are beginning to follow.

The final set of assumptions relates to the psychosocial trends likely to accompany and interact with these technoeconomic developments. Less familiar than the latter, they require more extended analysis.

The foregoing discussion of the technoeconomic assumptions was phrased in terms of the declining importance of work rather than of the increasing importance of leisure because of the crucial role that the former has always played in the psychological process of adapting to reality. Hitherto in human history, work has been directly and indirectly linked with satisfaction of the most powerful instinctual drives for nourishment and protection; it has been the principal means whereby the individual learns to obtain for himself food, clothing, shelter and usually at least a minimum sense of social and personal security. Thus, the necessity to work, physically or mentally, the concomitant need to acquire by imitation and training the personal characteristics and the manual or intellectual skills requisite for it in the prevailing sociocultural conditions, and the discipline of the work process itself have always been among the most important reality-adapting pressures, direct and indirect, immediate and delayed, in personality formation and maturation.

In his only major work devoted primarily to sociocultural analysis, *Civilization and Its Discontents* (1930), Freud wrote that "the life of human beings in common . . . had a twofold foundation, i.e. the compulsion to work, created by external necessity, and the power of love, . . ."[4] Earlier in the same essay, Freud explained:

> . . . work has a greater effect than any other technique of living in the direction of binding the individual more closely to reality. . . . Work is no less valuable for the opportunity it and the human relations connected with it provide for a very considerable discharge of libidinal component impulses, narcissistic, aggressive and even erotic, than because it is indispensable for subsistence and justifies existence in a society.

Freud went on to stress the anomaly that, at the same time,

> . . . as a path to happiness work is not valued very highly by men. They do not run after it as they do after other opportunities for gratification. The great majority work only when forced by necessity, and this natural human aversion to work gives rise to the most difficult social problems.[5]

The projection assumes, therefore, that work would cease to be a nearly universal necessity; that a majority of the population would have to work, if at all, only a few hours a day not simply to survive but to enjoy a reasonably satisfactory and even a rising standard of living. People would no longer experience from earliest childhood the gradually increasing pressure of knowing that they must sooner or later earn a living in competition with one another and that, before then, they must learn the behavioral norms and the occupational skills required to do so. In such circumstances, to what other kinds of comparably powerful positive and negative reality-adapting sanctions would people be subjected? Or, to use Freudian terms, how, in the absence of the necessity of work, would the reality principle (that is, the opportunities for and limitations on individual satisfactions existing in the natural and social environments) permit, postpone and prevent gratification of the egoistic drives comprised in the pleasure principle (that is, the effort to reduce the inner tensions and pain caused by failure to meet instinctual physical and psychological needs)? An answer to these questions is attempted in the next section. But, before that, certain of the reality-adapting characteristics of work need to be more fully explored.

Work may be enjoyable; as Freud pointed out, it may be the means for realizing other imperative gratifications besides subsistence and social acceptance. Indeed, for some people, especially the members of the various kinds of elite groups, work may be one of the most important forms of satisfying aggressive and erotic drives directly through the exercise of power or by sublimating them in creative intellectual and artistic activities. It may also be a means of gratifying narcissistic needs in the different types of display, homage and conspicuous consumption accompanying certain occupational roles. Pleasurable or not, however, work requires the expenditure of physical or mental energy in a recurrent systematized manner and has always constituted the largest and most important activity of the waking hours. Moreover, there is a crucial negative sanction involved that, unlike the positive sanction of the Protestant ethic, has hitherto been universal and immemorial. He who does not work physically or mentally in the ways prescribed by his society, class, group, family, shall not eat. Fear of losing the material and social necessities of life or of a privileged social position is the compulsive element that deprives people of the option of whether to work or not, regardless of whether they may be free to choose the specific occupations by which they will conform to the mandatory requirement. It is this negative sanction in all types of societies that differentiates work from play and from avocational activities in their many forms, which may otherwise be identical, equally arduous, and as regularized and time-consuming.[6]

Work and its associated social relationships—even in the broad definition of the term used here to include continuing full-time artistic, scientific and other intellectual activities—are not, of course, the only manifestations of the reality principle. Other kinds of individual and interpersonal experiences help to shape adaptation to the external environment by inculcating the willingness and ability to reduce, delay,

sublimate or repress many egoistic drives seeking gratification. Nonetheless, under this definition, there has never been a ruling elite either
in Western civilization or in any other that did not have to *work at
ruling:* that is, that did not feel compelled to train for and engage in
reasonably systematic and continuous activities of a military, religious,
judicial, political, administrative or other managerial nature. Even nonruling leisured elites—such as the Kyoto aristocracy during the centuries after the Heian period—if they were to survive for more than a
generation or two, had to evolve more or less elaborate life-styles in
which recurrent, systematized participation in love, sport, literary and
artistic pursuits, and religious and court ceremonials became mandatory, thereby fulfilling in the same compulsory fashion the reality-
adapting functions of work.

Suppose, then, that the age-old sentence passed on Adam after his
expulsion from the work-free plenty of the Garden of Eden—"in the
sweat of thy face shalt thou eat bread"—were to be repealed; that the
preponderant importance of and the compulsive element in work were
to decline and it were to become more and more of a subsidiary or
voluntary activity for a substantial portion of the population. What
changes in values, attitudes and behavioral norms would have to be
involved? Already in American society, less and less moral stigma is
being attached to not working among the people generally, if not to the
same extent among the elite groups. Indeed, shorter workdays and
weeks, longer vacations, earlier retirement, the proliferation of recreational products and facilities, the growth of leisure communities, and
other similar developments are increasingly regarded as desirable. By
the early 1970s, the absence of mass unemployment like that of the
1930s, and the rise in living standards for the majority of the population
were begining to offset the effects on children and young people of parental and other pressures, so marked during the postwar period, for high
educational achievement as the prerequisite to successful work careers.
Another attitudinal change is the perceptible decline in anxiety about
economic security in the successive age groups reaching maturity, even
when, as in years of little or no economic growth, there are difficulties
in obtaining first jobs. In the coming decades, these changes would
steadily weaken the sense among the people generally, if not among
the elites, that work and training for it were of such overwhelming
necessity that other life activities had to be rigorously subordinated to
them. Over the long term, such a trend would help to undermine two
of the most important ways in which work fulfills its reality-adapting
function: the effects of prospective and actual occupational roles on

formation of the sense of identity, and of success in them on self-confidence and self-respect.

These, then, are the main assumptions on which the median projection is based. The trends which they depict are either already manifest or incipient in American society, and there are reasons for believing that they could develop to become major determinative elements in it by the end of the century. If they do so, and if other important determinative factors do not nullify or significantly modify their effects, these trends will largely shape the characteristics of the 21st-century society. Some of their main implications are now worked out: for the constituent social groups and their life-styles in the next section, and for economic and political institutions and their relationships in the one following. However, these extrapolations do not constitute a comprehensive and detailed description of a future society, or even a more cursory *tour d'horizon* of all its principal aspects. Rather, they are only partially connected vistas of certain salient features of its sociocultural landscape.

The Median Projection: Social Groups and Their Life-Styles

One major consequence of the trends portrayed above would be a new differentiation of the population into three main groups. They would, of course, shade into one another, but each could be identified as a separate nexus in the social continuum, distinguished by different social roles and life-styles. I will designate them as the *technocratic and other elites,* the *leisured nonelites,* and the *dissenters and drop-outs* from both. The leisured nonelites would be the largest group, roughly 40 to 50 percent of the population; the other two would be about equal, each approximately 25 to 30 percent of the total.

The Technocratic and Other Elites

An economic system as productive, labor-saving and consumption-oriented as that implied by the foregoing assumptions would have to be highly specialized, internally and externally interdependent, and heavily reliant on continued scientific, technological and managerial progress at all levels. Scientific and technological research and development—physical, biological and social—and their institutional applications tend to be labor-intensive, as are the governmental, educational and other institutions needed to administer, train, preserve and mediate among the parts of so complex and intricately adjusted a society as that projected here. Even with the aid of the informational, computational and integrating capabilities of future computers, there-fore, ever-larger numbers of trained natural and social scientists, engi-neers, technicians, instructors, administrators and managers of all kinds would be required. Hence, the technocratic portion of the elite group would grow absolutely and relatively. This trend is already fore-shadowed in the numerous scientists, engineers, technicians and ad-ministrators required for research, development and application in the aerospace and other advanced industries; in the upgrading in auto-mated factories and plants of skilled workers into technicians; in the increase of civil servants in the middle and upper levels as government functions expand; and in the steadily rising percentages of young peo-ple obtaining higher education and advanced professional degrees. In the new period, therefore, the technocrats would comprise between two-thirds and three-quarters of the total elite group.

Most technocrats would continue to be strongly motivated toward work and personal achievement in it. The nature of their functions and of their competition for occupational advancement would incline most of them to a work day and a work week comparable to those now prev-alent among their contemporary forerunners, even though they might take substantially longer vacations and retire earlier. Indeed, for many technocratic elites, especially those at policy-making levels, overtime and weekend work would continue to be the norm. With the increasing professionalization of their occupations and the assurance of earning high salaries, few technocrats would be impelled to acquire great wealth for economic reasons, and it might also be morally frowned upon in consequence of the changes in attitudes induced by continuing criticism of pecuniary motives and predatory behavior. Nor would ownership of property persist as a major means for exercising power, which would more and more depend upon the competitive ability to reach the policy-making positions in the determinative institutions of the society.

Although no longer compelled to work by economic necessity, the large and growing group of technocratic elites would be constrained to do so by the nature of their socialization and acculturation in the family and the school and by other personality-forming influences, as well as for the satisfaction of egoistic drives obtained through the exercise of power, the pursuit of knowledge, and aesthetic expression. The largest portion of waking hours would be occupied during childhood and adolescence by the preparation for work, and during adulthood by work itself and the competition for occupational advancement, power and prestige. These activities would, therefore, continue to be major reality-adapting experiences. They would help to sustain and, in turn, would be perpetuated by values and behavioral norms that embody convictions of the worthiness of work as the means for fulfilling both the responsibility to employ reason and science for advancing the understanding of natural and social phenomena, and the obligation to apply this knowledge to social and individual improvement.

Most technocratic elites would be impelled to form families and to bear and rear children by their own internalized values and by the desire either to perpetuate themselves and their achievements or to compensate for their own unfulfilled aspirations through identification with their sons and daughters. Since vertical and horizontal mobility would be likely to persist in American society, the technocratic elites would be constantly augmented not only by natural increase but also by the addition of people from the other main groups, who were strongly enough motivated to acquire the necessary prolonged and difficult education, which would be freely available to all qualified students. A double process of natural and social selection would, therefore, at least maintain, and might even over the very long term significantly increase, the intellectual capabilites of the elites.

For the majority of children in elite-group families, as well as for those from nonelite families in which aspirations for higher social status would remain strong, work and the long, demanding preparation for it from childhood through adolescence until the late 20s would continue to be regarded as socially mandatory even though they would have lost the traditional sanctions of economic necessity and religious commandment. Hence, work as the most time-consuming and important waking activity, and the related pressure for competitive success in it, would still fulfill the major reality-adapting functions in the maturation process similar to those prevalent today and in the past. The example, instruction, approbation and disapproval of highly motivated parents, teachers, older siblings and peers would more than offset the contrary effects on most children of technocratic elites of affluence, economic security and the disintegrative factors, explained in the section on the decline-and-fall projection. This means that, in the coming decades, educational improvement would involve an effective compromise between the need for learning discipline and substantive curricula, on the one hand, and the desirability of fostering creativity and the enjoyment of childhood, on the other.

A majority of children of the technocratic elites would, therefore, eventually resolve their pre-teen and adolescent ambiguities and their conflicting impulses to identify with and differentiate themselves from parents and other paradigmatic figures. They would internalize exemplars, values and norms of behavior that would impel them to working careers, to striving for personal achievement, and to guilt over their failures. Indeed, in adulthood, the increasing accomplishments and ever-expanding capabilities of technocratic positivism, as well as their own personal successes, could lead to even greater confidence in their mastery over nature and society and would at least sustain, if not intensify, the sense of their redemptive mission to bring about what they believed would be further social progress and individual improvement.

The nontechnocratic minority of elite-group members would be quite heterogeneous. It would include politicians, the small and declining numbers of people with inherited wealth, the remnants of self-employed proprietors and of the older, less technical professionals, leading performers of all kinds in the large and varied entertainment industry, and the literary and artistic intellectuals. Except for the wealthy, they would be work-oriented, like the technocrats, although in the main their leisure-time activities would be similar to those of the leisured nonelites. However, the life-styles of many of the intellectuals and the entertainers would tend to resemble those of the dissenters.

The Leisured Nonelites

In accordance with the declining importance of work, the attitudes, motivations and life-style of the large nonelite portion of the population would gradually diverge markedly from those of the technocratic elites. Consisting of the equivalents of today's blue- and white-collar workers, the nonelites would be guaranteed adequate incomes, as then defined, for working a few hours a day. Differential supplementary wages, preferred residential privileges, and other incentives would probably be needed to assure that sufficient manpower would be available to meet the various labor requirements, especially for the less agreeable occupations. However, those who did not wish to work or were unable to do so would receive minimum incomes that would permit them to have a reasonably satisfactory standard of living, even if lower than that of the employed.

Successive age groups of the children of the nonelites would be increasingly subjected to the examples and instructions of parents, other adults, siblings and peers, for whom the necessity of work and the pressures for occupational achievement would be felt as less and less compelling. This shift in values and behavioral norms would be reflected in, and would in turn reinforce, changes in their education, which would tend toward the extreme of unstructured permissiveness. The other manifestations of the reality principle that would take the place of work and disciplined preparation for it would be analogous to those of the nonruling leisured elites of the past, and more or less similar to the activities that already fill the nonworking time of the great bulk of American families with incomes above the poverty level today.

They would include visiting and entertaining; dating, courtship and sex; tourism and outdoor recreation; active sports and hobbies of all kinds; spectator sports, gambling, TV, movies and other amusements; ceremonies and celebrations; and participation in local social clubs, political organizations and religious institutions. Such ways of interacting with external physical realities and with other persons would gradually be felt to be of greater, and eventually of preponderant, importance. These experiences would become more and more systematized, repetitive and time-consuming, and participation in them would be increasingly impelled by the threat of social disapproval or even ostracism. Intimations of the ways of life of the increasingly leisured nonelites and their children may be seen in those of contemporary retired middle-class couples still young enough to be active, and of adolescents "in the typical suburban high school world of sports, sports cars, girls, rock and roll, academic cheating and disparagement of intellectual accomplishment."[7]

While fulfilling reality-adapting functions, such activities would require much less rigorous and prolonged training and would more readily, immediately and directly gratify narcissistic, aggressive and erotic needs than would the maturation process and the life-style of the technocratic elites. Several consequences for personality formation of the nonelites would follow from these differences. The greater scope for and hedonistic nature of egoistic drives combined with the declining importance of internalized values fostering creative sublimation and the sense of social responsibility would substantially increase self-concern and self-indulgence and would weaken the ability to postpone or forgo immediate gratifications. Neither inherited social status, as in traditional hierarchical societies, nor significant occupational roles, as in contemporary Western societies, would contribute powerfully to the sense of identity. Their absence would reinforce the effects of the narrowing of distinctions between the sexes in intensifying feelings of ambiguity and alienation.

Thus, the way of living of the nonelites would not be without serious psychological difficulties and social problems. In contrast to the technocratic elites, successive generations of increasingly leisured nonelites would be less and less inclined to bear and rear children. The likelihood that safe, efficient and socially approved methods of contraception and abortion would be freely available would make it possible for them to refrain from doing so. This trend would reflect the gradually increasing reluctance of the self-concerned and self-indulgent nonelites to forgo direct and immediate gratifications and to assume the prolonged interpersonal responsibilities required of parenthood even in the absence of economic pressures. Moreover, since their accomplishments would largely be consumptive rather than creative and their desired satisfactions would be readily obtainable and repeatable, they would be much less impelled to perpetuate their achievements or compensate for their failures by identification with children. Instead, pets might become increasingly popular, for they would be less troublesome and demanding objects of affection.

Hence, as families became smaller and marriage partners were more frequently changed, the birthrate would be likely to decline among the leisured nonelites. Opportunities to join the technocratic elite would be available for those children who, whether by genetic inheritance or by atypical family and school experiences, were motivated to acquire the necessary professional education and to change their life-style. In consequence, although continuing to be the largest group in the society, the leisured nonelites would be a slowly declining proportion of the total population which, in the 21st century, would probably begin to recede from the peak reached during the latter decades of the 20th century.

Moreover, it is possible that, if the leisured nonelites were as disinclined to assume the main responsibility for rearing their children as the projection indicates, day nurseries, residential schools and similar institutions would become much more important than the family in the socialization process. If such a development were to occur, it would mean that peer groups and communal identifications and sanctions would play the major roles in personality formation. Although experiencing much less alienation and neurosis, institution-bred adults would tend to be conforming, incapable of intimate interpersonal relationships, and lacking in the emotional depth and complexities, usually attendant on family rearing, that are often sublimated in creative intellectual, artistic and literary activities.[8]

The Dissenting Groups

In the majority of the leisured nonelites, the problems arising from narcissistic and aggressive behavior and schizoid tendencies would not be so pronounced as to make them nonfunctional nor would prevailing moral and intellectual considerations motivate rejection by the society of their hedonistic way of life. However, for a minority of nonelite children, especially those who continued to be raised in families, the guarantee of economic security would reinforce the effects of the psychological factors described in the decline-and-fall projection in impelling the expression in various forms of feelings of alienation, meaninglessness and inadequate sense of identity. For these reasons, too, a minority of technocratic-elite children would experience the same reactions. Thus, young people alienated from rationalistic and hedonistic modes of living would continually be joining the dissenters and drop-outs already comprised in the third major group of the society.

Unlike the other two groups, however, the third would be highly heterogeneous, embracing a wide variety of life-styles that would share only the common characteristics of being different from those of, and more or less disapproved by, both the technocratic elites and the leisured nonelites. Much as they would frown upon dissident life-styles, however, the technocratic elites and leisured nonelites would be more tolerant of them than is the case today in consequence of the increased effectiveness of humanistic values and the absence of serious economic pressures.

A major social function of the third group would be, therefore, to maintain a considerable degree of diversity and decentralization of initiative that could continue to stimulate and enrich the society as a whole. In part, the third group would overlap with the nontechnocratic portion of the elites engaged in artistic and literary activities and in mass entertainment. The diverse reactions of the third main group to the other two dominant life-styles would cover a broad spectrum ranging from those largely determined by unconscious psychological processes to those mainly shaped by rational considerations.

In the first category would be the many kinds of neurotic and psychotic behavior fostered by the weakening of social restraints and inner repressions. The lessened sense of external reality and the need for more vivid and dramatic sensations would impel greater resort to fantasies and other inner experiences induced by drugs—no longer necessarily harmful—self-hypnosis, group hysteria, and other means. The inability to cope with either the complex rationalized way of life of the technocratic elites or the compulsively hedonistic way of living of the leisured nonelites would lead to individualistic withdrawals and relational simplifications. Expressing protoschizoid or protoparanoid processes, these reactions would manifest themselves in different kinds of eccentric, but not necessarily dysfunctional, behavior patterns.

In the second category would be the more or less rational efforts to organize mass protest movements of various kinds, and small philadelphic communities and new family-type groups. The mass movements are discussed in the next section. Whatever their specific designs, the small communities would all have a similar aim: realization of humanistic values and individual potentialities in ways that avoided both the impersonal rationalism and *anomie* of large-scale technocratic organizations and the dehumanizing effects of the leisured nonelites' inadequate repression and inability to sublimate narcissistic, aggressive and erotic drives in creative activities. Efforts to achieve this objective would probably be more successful than those of Fourier, Saint-Simon, Owen and today's humanistic utopians and neoanarchists. The enormous productivity of the technocratic society and the guarantee of a minimum income to all its members would exempt such 21st-century communities from the economic difficulties that were so often fatal to their 19th- and 20th-century forerunners. Moreover, they might derive supplementary income not only from growing and making additional materials and goods for their own use but also by meeting the elite-group demand for the handicrafted, individualized and high-quality products and services that could not be provided by automated factories and large service organizations geared to mass production and consumption.

Nevertheless, the likelihood that these small dissenting communities could fulfill an aesthetically useful economic role would not alter the fact that their ability to do so would depend upon the vast, intricate, rationalized economic system of the technocrats, which could alone assure the productivity required for the high living standards of the society as a whole. Thus, many of their members would feel the gnawing frustration of the paradox that the continued existence and success of their own communities would be contingent upon those of the hated technocratic order they aspired to replace.

The economic viability of small philadelphic communities and deviant family-type units would enable many of them to persist for several generations. They might, therefore, be able to develop the collective sense of identity and high morale that characterize some of the existing communities of Mennonites and other radical Protestant sects, as well as the Israeli *kibbutzim.* If, as in the case of the latter, they were also to practice communal child-rearing from earliest infancy, the personality type likely to predominate eventually would be conscientious, hard-working and serious but also limited, conforming and static. As Bruno Bettelheim concluded of the second-generation children of the *kibbutzim:*

> They feel no need to push ahead, but neither do they have the impulse to push anyone down. While such people do not create science or art, are neither leaders nor great philosophers nor innovators, maybe it is they who are the salt of the earth without whom no society can endure.[9]

Again paradoxically, such small communities originally founded to replace the rejected technocratic society might in time become its unwitting supports. This possibility suggests that, if communal child-rearing became widespread among them, the third major group could only fulfill its social role of serving as the source of creative diversity and humanistic dissent in the rationalized and hedonistic future society by continuous addition of new dissidents and drop-outs from the other two groups rather than by its own natural increase.

The Median Projection: Some Aspects of Institutions and Relationships

Based on the assumptions and the social structure and life-styles sketched in the preceding sections, it would be possible to work out a comprehensive description of the kind of society and culture they imply. Numerous detailed constructions of this kind have been undertaken by political philosophers, utopian novelists and science-fiction writers. However, for reasons already indicated, I shall limit myself to commenting briefly on several implications that seem to me especially relevant to the two extreme projections presented earlier in this chapter. The purpose is to show how the median projection incorporates in less developed forms certain of the different trends characteristic of each of them, and hence can be regarded as occupying a middle position in the range between them.

Changes in Public-Private Relations

With its satisfactory functioning and further development so greatly dependent upon advances in the natural and social sciences and their applied technologies, where would the locus of power be in the society depicted in the median projection? Would it shift to the universities and independent research institutes or would it remain, as hitherto, in government agencies and economic organizations? Certainly, the universities and research institutes would become even more important and influential than they are today for the reason just given. But, as institutions, they are not adaptable to the direct exercise of political and economic power in the society. They do not themselves possess sanctions with which to enforce their will; their organizational structures are too loose to be focused continuously on achieving external objectives; and their professional staffs are too mobile to cooperate for long enough periods for such concerted efforts to be generally effective. Therefore, it seems likely that, although the society would be much more dependent than is now the case on both the teaching and the research functions of the universities, their correspondingly greater role in policy making and sociocultural change would nevertheless continue to be effectuated indirectly through their influence on the attitudes and ideas of the technocrats in other institutions and organizations, public and private, actively engaged in policy formulation and implementation.

The main change affecting the exercise of power would probably be much greater blurring of the distinction between the government and the private organizations comprised in the economy and the other major institutional systems than now exists. This does not mean that the 21st-century society would be socialist in the conventional sense of the term. True, private ownership *per se* of the means of production would be even less important than it is today in determining who would control large corporations—which would be completely run by professionally trained managers and technicians, that is, by the technocrats. Rather, the interweaving of public and private organizations would be much more significant. It would be fostered not only by the close cooperation required for successful *macro*-economic management in highly specialized and internally and externally interdependent systems, but also by the similar professionalization of the technocratic elites in rationalized public and private institutions and their mobility within and among them.

The result would be the gradual narrowing of the "arms-length" relationship, characteristic of the liberal order, that has hitherto prevailed in the United States. Moreover, multinational enterprises would be increasingly impelled to think of themselves and to act as though they were like governmental institutions in consequence of their wider-than-national horizons and options and the cosmopolitanization of their managerial and technical personnel. The blurring of the distinction between public and private would be analogous to—not identical with—the relationship that existed in patrimonial societies, with multinational corporations playing roles equivalent to those of the great quasi-sovereign international trading and financial companies of the 16th to the 18th centuries. However, multinational enterprises would not be as independent in the median projection as they are assumed to be in the new feudalism of the decline-and-fall projection.

The New Pluralism

Nor in the median projection would the blurring of the distinction between public and private be accompanied by, and hence it could not reinforce, the other disintegrative trends included in the decline-and-fall projection. Indeed, this difference would reflect, and in turn help to preserve, the society's continuing integration and dynamism. As already today, the dispersal and growth of power in large corporations, trade unions and other private institutions, as well as in local and regional governmental agencies, would tend to hold in check the centralization and authoritarianism inherent in the technocratic society without impairing its effective functioning. Also, continuing competition—as well as cooperation—among public and private agencies would help to sustain innovation and managerial vigor.

Unlike the trend in the decline-and-fall projection, the enhanced importance both of private and of decentralized official organizations would not be accompanied by the growth of a relationship of feudalistic dependence between them and their employees, suppliers, distributors, etc. Physical danger would not be nearly as great in the median projection, and economic security would be guaranteed. With the declining significance of work and the smaller portion of their waking time spent in it, the leisured nonelites would feel much less involved in, and would have fewer opportunities and incentives to identify with, the public and private institutions employing them. In contrast, the technocratic elites, despite their continuing job mobility, would in the future identify more strongly than today with the institutions in which they worked because these organizations would be much more completely under their control—that is, they would be more fully technocratic in character.

For the leisured nonelites, other types of affiliations would become more important than the ties to their employing organizations. They would include not only trade unions, which would continue to protect the economic interests of the working nonelites, but also the institutions associated with their increasingly significant nonworking activities— sports and social clubs, entertainment centers, religious congregations, fraternal orders, residential and neighborhood organizations, local government agencies, and others. The nonelites would value these institutions both for the benefits they provided and for the satisfactions obtained from the process of participation in them. Identification with such organizations would help to strengthen their inadequate senses of personal identity, and involvement in membership meetings, committee assignments, local election campaigns, ceremonies and similar activities would permit gratification of the egoistic drives formerly satisfied by factory or office work.

Similarly, the technocratic elites in government and the economy would attach greater importance than they do today to membership in professional societies and to continuing ties with universities and independent research organizations. These institutions would provide the technocratic elites with a substantial portion of the periodic retraining —the continuous education throughout their professional careers—that would be necessitated by the growing specialization of science and technology and the constant expansion of knowledge and techniques. Possessing in this way the keys to occupational status and personal advancement, professional organizations and academic institutions would become even more like guilds than they are now, defining the standards and grades in their disciplines, protecting their members' interests, and seeking to enhance their own power and influence in the society.

These developments would significantly change the pluralistic character of contemporary Atlantic societies. Pluralism would not disintegrate into particularism, as in the decline-and-fall projection. But, the competing and bargaining interest groups and organizations comprising the 21st-century society would be different from those that predominate today. True, nationwide business councils, trade associations and federations of labor unions would probably persist, and competitive interactions among them would, therefore, continue. However, their political influence would decline because individual corporations and trade unions would be large and powerful enough to be less dependent on their services. The other types of interest groups and their representative organizations noted above would become more important, especially the professional societies of the different kinds of scientists, technicians, managers and other technocrats. Even the organizations of the leisured nonelites would be more active politically, not only in local affairs, which their members would feel were more important for maintaining and improving their nonworking activities, but through countrywide associations at the national level as well.

National Politics and Policy Making

Traditional American conservatism with respect to political institutions would be likely to preserve the existing forms of representative government even though the popularly elected officials would tend more and more to be technocrats with the positivistic urge to rationalize the legislative bodies and executive agencies in which they served. However, more important than the relative inefficiency of obsolescent government agencies in complicating rule by the technocrats would be their own increasingly limited capacity to provide effective popular political leadership. On the one hand, they would become more adept at manipulating public opinion as the applied capabilities of social psychology and sociology were developed. On the other hand, their growing rationality and impersonalism would restrict their ability to evoke the emotional responses and loyalties that are essential for holding popular attention and support.

In consequence, there would be a continuing role for the traditional type of politician, whose leadership capabilities depend primarily upon the requisite personality traits rather than on technical training and knowledge. The persisting criticism of the technocratic society by the dissenters, and the American political custom of local reformist initiatives would also help to produce political leaders capable both of winning popular support and of cooperating effectively with the ruling technocratic elites. However, the life-style of the leisured nonelites would incline them to favor politicians drawn from the occupational backgrounds of greatest and most direct importance to them—actors and other entertainers, officials of their own nonworking-time organizations, and spectacular personalities, rather than the lawyers and small businessmen comprising the majority of politicians today. On the one hand, this trend would make the society even more dependent upon the knowledge and skills of the technocrats. On the other hand, it would mean that the control of the technocrats and of the responsible politicians would be continually threatened by—and probably intermittently lost to—demagogic leaders, whose own tendencies toward megalomania and paranoia would give them the psychic energy and flamboyant appeal needed to organize the mass movements discussed below.

National policy making would ordinarily be even more the concern of the technocratic elites than it has already become, except in unusual circumstances when popular anxieties would be aroused. The median projection assumes that many contemporary political issues relating to the distribution of income and the allocation of resources would have minor, if any, importance in the 21st-century society. However, other choices among goals and priorities, domestic and foreign, would be highly controversial because resources are never unlimited and differ-ent institutions and groups of technocrats, nonelites and dissenters would conceive both their own and the social interest in conflicting ways. Also, a large and growing portion of national politics would con-sist of disputes among the various technocratic elites regarding the most effective means for achieving agreed-upon objectives; scientific controversies; conflicts between humanistic criteria and efficiency cri-teria in policy making; and bureaucratic rivalries and factional and personal struggles over power and prestige within and among public and private organizations. That contentions of these kinds could con-stitute much of the substance of national politics in a society increas-ingly governed by technocratic elites is indicated by the rational and emotional importance attached to such issues today in the internal "politics" of large corporations, universities and other institutions staffed by their contemporary counterparts. Finally, demagogic poli-ticians and activist dissenters would periodically interject divisive and distracting or radically transforming issues into politics that would tax the political capabilities of the technocrats and divert resources to objectives of which they deeply disapproved.

Social Dissension and Mass Movements

These political developments would reflect the steadily widening gap between the ways of thinking and acting of the technocrats and those of the leisured nonelites and the dissenters. This trend would be fos-tered by the technocrats' need for a system of education intellectually much richer, more structured and prolonged than that of the other groups, even though the differences might be veiled by euphemisms and subterfuges. The growing cultural gap between the main social groups would constitute a serious discontinuity in the 21st-century so-ciety, likely to aggravate all of its other problems. It would lead to recurrent misperceptions and misconceptions on the part of techno-cratic policy makers regarding the concerns and expectations of the people, with consequent frustration on both sides, periodic revivals of popular participation in national politics often under demagogic or messianic leaders, resurgence of the endemic American anti-intellectualism, and occasional social unrest and violence. In the intervals of such reactions, the cultural gap would make the manipulation of public opinion by the technocrats an uncertain process, despite their technical skills, and it would be a continuing stimulus to and target for the attacks of the dissenters.

The characteristics of the 21st-century society have implications for the social and intellectual modes of expressing disaffection and radical dissent. For example, crime against property mainly induced by economic need would tend to disappear in so affluent a society, with its guaranteed incomes for all and tolerance of deviant and eccentric lifestyles, including drug-taking—which scientific advances might make no longer physiologically harmful. In contrast, crime against persons primarily impelled by psychopathological difficulties would increase, especially among the leisured nonelites and the dissenters and drop-outs, owing to their proneness to immediate gratification of aggressive feelings and their lesser capacity for self-control.

Fundamental dissent from the technocratic society would not take the form of organized revolutionary movements of a predominantly politicoeconomic character. Exploited discontented workers and farmers would not exist to provide a "mass basis" for such movements, and the depressed minorities would by then be sharing equitably in the benefits of the society. Neither could radically dissenting students and other young people fulfill this role because they would not normally be members of institutions vital to the day-to-day functioning of the society, and their participation in schools, universities, youth groups and similar organizations is in any case always temporary—for, alas, they soon cease to be young and, eventually, students. Moreover, the dissenters and drop-outs comprised in the heterogeneous third major social group would tend by nature to be too individualistic and un-disciplined to constitute a sufficiently stable and reliable basis for continuing mass revolutionary movements.

Indeed, it is possible that revolutionary efforts would tend much more than those of today to be religious rather than politicoeconomic in their modes of expression and organization. At one extreme, the need for sense-saturation and the search for meaningfulness and identity would be met by the spread of ecstatic and even orgiastic religious cults, most more or less Christian but some satanic or mystical in their theologies and rituals of personal salvation. At the other extreme, there could be revivals of Christian and adaptations of Buddhist and other Oriental types of monasticism and anchoretism (i.e., living as a hermit) —of communal and individual withdrawal from society as the path to salvation—for those who required austere, contemplative and closely controlled ways of living physically separated from both the highly rationalized and the excessively hedonistic life-styles of the other two main social groups. The former would tend to develop from the evangelical and pentecostal Protestant sects, the latter from Catholicism and the more rationalistic Protestant churches.

Between these two extremes, both branches of Christianity could provide doctrine, ritual and organization for mass movements of a social-redemptive character that might from time to time crystallize around charismatic messiahs. Such religious protest movements would tend to be puritanical, fundamentalist, authoritarian and anti-intellectual, promising social renovation, as well as personal salvation, through suppression of the arrogant rationalism of the soulless technocrats, of the sinful hedonism of the leisured nonelites, and of the anxiety-provoking eccentricities of the dissenters. Because the prevailing life-styles would be more or less unsatisfying to a substantial minority of the participants in them, religious or quasi-religious messianic movements would be able to attract—and near their peak to coerce—large, if temporary, followings from all three main groups. During the years of their greatest popular support, these movements—or rather their messianic leaders—would have substantial political importance. And, the misperceptions and misunderstandings arising from their own rationalistic secularism would make it especially difficult for the ruling technocrats to cope effectively with such popular religious upsurges.

Implications for Intersocietal Relations

The kind of 21st-century portrayed in the median projection would be consistent with the projection in Chapter IV of the way in which the present period of world politics might evolve. In a fully developed technocratic society, in which humanistic values also helped significantly to shape self-conceptions and behavior, both the positivistic conviction and the sense of redemptive mission would continue to be powerful elements in the dramatic design of the elites, motivating an activist approach to problems and opportunities abroad as well as at home. Thus, the United States would maintain its superpower role during the remaining decades of the century and thereafter would continue to be a principal participant in the international system.

The median projection would also be consistent with the most probable of the long-term developments envisaged for the Atlantic regional system in Chapters V and VI. The blurring of the distinction between public and private organizations and the greater importance and independence of multinational enterprises would coincide with growing transnational integration at both governmental and private levels. The different institutions developing closer Atlantic-wide ties and interdependencies would at the same time be attracting the loyalties of the technocratic elites away from the nation-state *per se* because they would be increasingly significant foci for the senses of identity and purpose of their policy-making and technical personnel. The result in the 21st century would be the cosmopolitanizing of the technocratic elites, a development also sanctioned by the universalism of their rationalist conceptual framework and further fostered by their more and more similar behavioral norms. In contrast, because the attention of the other two major groups would most often be focused predominantly on their immediate personal concerns and local affairs, their ways of thinking and acting would be parochial. But, their lack of interest in regional and world affairs would probably inhibit them from being actively xenophobic, and hence interfering with the internationalist policies of the technocrats, except intermittently when mass movements under demagogic leaders might temporarily revive a strong national consciousness. Thus, the existing predominantly vertical focusing of loyalty and concern upward to national governments would be counterbalanced by an equally significant horizontal focusing on the increasingly important transnational private and governmental institutions of the Western technocratic societies. In this way, the period of the new nationalism would be transformed into the period of the *new pluralism* in regional relationships, if not on a worldwide scale.[10]

Neither the Fall nor the Millennium

The Atlantic societies characterized by the new pluralism would be neither declining toward their inevitable fall nor rapidly approaching the utopias of social perfection or personal fulfillment. Inherent in the life-styles and relationships outlined in the foregoing pages are serious social problems and individual difficulties no less perplexing than, although different from, those of the period of the new nationalism. Nevertheless, this projection tends toward the optimistic end of the possible range. Strictly speaking, it is not a median but an average, reflecting my judgment that the balance of trends favors social integration rather than disintegration. Different assessments, especially of the extent of success during the remaining decades of the present century in dealing with the disintegrative forces, would lead to projections closer to the pessimistic end of the range. They would reflect not only the failure to integrate the deprived groups more equitably into the society but also the maintenance of hours and conditions of work like those of today. If the nature and duration of mass production in factories and offices is unchanged, the already incipient trends toward boredom, depersonalization and declining conscientiousness and quality would become manifest, reinforcing the other disintegrative trends. The outcome would then be more like the new feudalism of the decline-and-fall projection than the new pluralism, with its reduced importance of work.

If the median projection is likely to be rejected by the pessimists as too optimistic, it will be equally unsatisfactory to the technological futurists and the humanistic utopians.

To project a possible 21st-century society and culture embodying ever more far-reaching advances of science and technology, yet in which sovereign reason would still be a limited and not an absolute monarch, is to deny one of the earliest and most cherished convictions of technocratic positivism. And, to imply further that the operational effectiveness and creative potentialities of a fully developed technocratic society would be not only threatened by but also dependent upon nonrational processes is to add the insult of contradiction to the injury of refutation. Such a projection is especially unwelcome to American positivists, whose cultural heritage, professional training and peer-group conformism continually reinforce the conviction of their imminent mastery over nature and society through the progress of science and the power of reason.

At the same time, to project a possible 21st-century society and culture characterized by continuing high consumption, an increasingly hedonistic and narcissistic life-style for the largest group of the population, and persisting psychological problems and personal unhappiness is to disappoint those who envisage that mankind will soon free itself from the "tyranny of things" to realize its potentialities for spiritual, aesthetic and intellectual development. Yet, technological advances, the economy of abundance, and the decline of work are not likely to make all, or even many, of us gurus, artists or philosophers. Indeed, that a majority of human beings would in the foreseeable future pursue, or even be much interested in, the discovery of the transcendental, the experience of the beautiful, or the life of the mind is improbable without a further evolutionary change—in the Darwinian sense—in the nature of man. Bernard Shaw was probably right in envisioning such a "metabiological" transformation as occurring only in the final play of his futurist pentology, *Back to Methuselah,* which he entitled "As Far as Thought Can Reach" and dated in 31,920 A.D.!

The Future of Society and the Future of the Social Sciences

The effectiveness of technocratic elites in advancing knowledge and helping to shape the future depends directly upon their willingness and ability to recognize and take account of the complex and contradictory interactions among egoistic drives, rational interests, and the particular ways in which the expression of both is shaped by institutions and values. Because the harm that could result in the increasingly techno-cratic society from rationalist over-confidence would be exceeded only by that perpetrated by ignorance or fanaticism, I believe it would be useful to discuss some of the constraints involved in policy making and implementation and the possible developments in the social sciences that might ease them over the longer term.

Max Weber had profound insights not only into the history and con-temporary nature of society and culture but also into the eventual out-come of the process of continued rationalization in large-scale political and economic organizations and professionalization of their personnel. On the one hand, he stressed their essentiality for maintaining the rising productivity of a complex, interdependent socioeconomic system. On the other hand, he feared that increasing rationalization would even-tually crush human freedom in the "iron cage" of depersonalization and regimentation, and overwhelm human creativity by the material-istic self-indulgence which the resulting affluence would permit on a mass scale for the first time in human history. In the conclusion to his best-known work, *The Protestant Ethic and the Spirit of Capitalism,* published in 1904–5, Weber wrote:

> No one knows who will live in this cage in the future, or whether at the end of this tremendous development entirely new prophets will arise, or there will be a great rebirth of old ideas and ideals, or, if neither, mechanized petrification, embellished with a sort of con-vulsive self-importance. For of the last stage of this cultural develop-ment, it might well be truly said: "Specialists without spirit, sensual-ists without heart; this nullity imagines that it has attained a level of civilization never before achieved."[11]

Weber's scorn was matched by his pessimism, since he believed that humanistic values, the only countervailing power he could discern in such an overly rationalized society, would be too weak to offset its deadening effects on the human spirit.

Sigmund Freud, too, was pessimistic about the future not because he believed that reason would go too far but because he was afraid it wouldn't go far enough in organizing and controlling society and the individual. To Freud, civilization is a human invention for repressing, channeling and transforming the egoistic drives, whose unrestricted efforts to obtain satisfaction would otherwise result in mutual and self destruction. While recognizing that repression is a major cause of neurosis and devising a therapeutic technique for mitigating its pathological effects, Freud also feared the free expression of the instinctual impulses. "What an overwhelming obstacle to civilization aggression must be," he wrote, "if the defense against it can cause as much misery as aggression itself!"[12] For him, reason was the countervailing power but, like Weber's humanistic ideals, not strong enough to harmonize the blind force of the pleasure principle with the harsh restraints of the reality principle.

Contrary to the expectations both of technocratic positivists and of their humanistic critics, this dichotomy seems to be more naked and extreme the more rationalized and reformed the society becomes. In Freudian terms, the triumph of secondary process (conscious rational thinking) appears to bring with it the periodic unleashing of primary process (unconscious psychic activity expressing instinctual drives). Already there are intimations in our own current experiences that the more ethically concerned and reasonable—that is, the more just and tolerant—the society is, the more unreasonably dissent from it is manifested. Much criticism has been directed against the noncommunist New Left and other radical student and youth groups for their lack of a program, of an alternative viable system of values and institutions to replace those they so scornfully and violently reject. Yet, in a reasonably well-ordered and increasingly more equitable society, it is more and more difficult to devise a competing principle of order and meaning with which to organize a new and different society.[13] Hence, the dissent of the more intelligent and highly motivated youth tends to be radical, in the literal sense of the word, a root-and-branch critique that, *at its best*—like Paul Tillich's "Protestant principle"—justifies itself not by presenting practicable alternatives but by compelling existing values and institutions to justify themselves or perish. Moreover, because the legitimate means of protest, as defined by modern Western society, are themselves manifestations of its tolerance and justice, they, too, must be rejected by dissenters. The psychological pressures to express narcissistic and aggressive impulses are validated by the social requirements for meaningful dissent, with the result that it tends to be violent, nihilistic and episodic.

The danger is that violence and nihilism—as well as the different, less immediately destructive forms of freely gratifying egoistic drives advocated by Herbert Marcuse, Norman Brown, Timothy Leary and other older mentors of youthful dissent—may be elevated from means of expression to ends in themselves. If, when reason and humanistic values fail, it is the impossibility of completely controlling primary process that keeps human freedom and creativity alive even under the most repressive totalitarian regimes, it is nevertheless equally true that the gratifications sought by egoistic drives can be provided only in a society, not in a Hobbesian state of nature. And, a society is, by definition, a system of organized restraints and orderly ways of rationing satisfactions and controlling the forms in which they are realized. Hence, radical dissent must not destroy society; its only justifiable function is, when the need for reform is imperatively felt and legitimate means of protest are ineffective, to force a fundamental reevaluation by the society of its established institutions, relationships and norms of behavior. For, if such self-judgments are in fact to remedy injustices, they can only be made and carried out by the method of reason in accordance with the standard set by humanistic values. The well-known tragedy of violent social revolutions is that they more or less negate both the ends and the means that alone can justify them. Yet, it may be only another paradox of existence that humanistic values and the rule of reason are made operationally effective in human affairs as much by the motivational power of primary process as by the need to correct its unrealism and control its destructiveness.

In seeking to understand these paradoxical interdependencies and to devise remedies for coping with the difficulties they generate, American social scientists are impeded by the biases bred into them by their culture and the limitations of the disciplines in which they have been trained. As explained in Chapter III, technocratic policy makers in governmental and private institutions tend to ignore or minimize the roles of sociocultural and psychological factors in analyzing and prescribing for the problems with which they are confronted.[14] When it sooner or later becomes apparent that their initial simplistic economic approaches—as in the poverty program and the foreign-aid program— are falling seriously short of expected results, their pragmatism leads them to recognize the importance of noneconomic elements and nonrational processes. If, then, they turn for help to sociologists, political scientists, psychologists and others, they are often drawn to those who advocate grandiose schemes of social engineering or purely technical panaceas that are supposed to accomplish major social transformations. And, finally, when the complexities and contradictions of human nature and society are taken into account, there is a tendency to conceal their essential significance in rationalistic euphemisms, such as "trade-off" and "second best," that give no sense of the force of the frustrated egoistic drives and interests involved and, hence, of their disruptive or destructive social and individual consequences.

One way to achieve the better perception of the ambivalences and constraints of reality that could more quickly bring down to earth the enthusiasms of redemptive activism would be through changes in the conventional social-science disciplines and their respective specialized subdivisions. If the functions of the social sciences as well as their location were solely academic, the probability of this development would be low. It is more likely to be high, however, precisely because positivistic social scientists generally—like Marx's philosophers—are concerned not simply to understand social institutions and processes but also to change them. Therefore, just as the existing academic disciplines replaced those of the medieval universities when the function of knowledge shifted from "justify[ing] the ways of God to men" to discovering the laws of nature and society, so the contemporary divisions and specializations in the social sciences will probably be reintegrated in new ways by the felt need to foresee and shape the developing future.

The difficulty is that the panorama perceived by and of interest to natural and social scientists in a technocratic society is already enormously large and full of detail and is rapidly expanding both extensively and intensively. This broadening and deepening of knowledge have led perforce to more numerous and narrower specializations, which are further entrenched by the collegial exclusivism and professional conformism inherent in the academic institutions of Western societies. However, the "trained incapacity"—to borrow Veblen's term—of such specialists to deal with complex real-life problems is already beginning to be recognized and efforts to overcome it are being made by means of multidisciplinary teams and integrating methodologies, especially systems analysis. Both of these new approaches are promising but, as presently practiced, they have serious weaknesses which reflect the fact that they were originally developed to meet engineering rather than social-science needs.

The use of mixed teams of engineering specialists in designing complex products and research or production processes is successful because the subject matter is always a more or less homogeneous continuum of physical interrelationships, and the specialized fields of knowledge involved are complementary and usually overlap sufficiently to minimize communication difficulties. Hence, such teams can generate a capability that is qualitatively greater than the sum of its parts. In contrast, most social problems are much more heterogeneous, comprised of different institutional, cultural and psychological factors and interactions. In most cases, these complex processes are not nearly as well understood as are physical phenomena; the specialized knowledge of social scientists is often discontinuous; and concepts for intercommunication among different disciplines have not yet been developed in many fields. In consequence, mixed teams of social-science specialists usually fail to produce an integrated approach to real-life problems—and often also to scholarly efforts undertaken by multidisciplinary groups.

Systems analysis, too, was devised for dealing with complicated engineering problems in research and production. As a methodology, it is by nature well suited to the analysis of most types of interactions in many kinds of systems, physical and social. The difficulty lies in applying it. It is not enough to be trained in the use of the technique *per se;* also essential is knowledge of the subject matter to which it is being applied. Both conditions are generally met by the engineers using systems analysis in the solution of engineering problems. They are not usually met by the engineers, mathematicians, econometricians and others trained in the physical sciences or in abstract methodologies who have hitherto taken the lead in trying to apply systems analysis to social-science problems and national-policy issues.

The need for professional knowledge of the social sciences in order to use systems analysis fruitfully in dealing with such problems is underscored by the limitations of the social sciences themselves. In most areas of social experience, understanding is not yet sufficiently advanced to formulate the equations for representing the highly intricate processes involved. And, even when models can be constructed, data are often lacking for filling in many of their terms, and many others may be unquantifiable. Hence, approximations and estimations—quantitative and nonquantitative—are usually required, and they can best be made by people whose professional training gives them an informed sense of the fitness of assumptions, hypotheses and conclusions.

The growing need for professionals with such capabilities is likely to lead sooner or later not only to the regrouping and redivision of the conventional social-science disciplines and specializations but, more important, to the development of a *new kind of scientific integrator or generalist*. Hitherto, the social scientists—myself included—who wished to integrate concepts and data from different subject fields in order to devise more effective policy prescriptions have had to be self-motivated and self-trained, with inevitably serious gaps in the information they require and limited capacity to use the growing variety of sophisticated analytical techniques appropriate to the different disciplines. But, the emerging demand for the services of such integrators is now beginning to impel universities to consider devising graduate curricula specifically designed to provide them with the knowledge and technical skills they need. As this trend develops, it will be analogous to that which occurred in American universities in the decades from 1890 to 1920 when, as explained in Chapter III, the existing graduate schools and curricula were established in response to the demands for professionalized personnel initially from business corporations and later from governmental and other institutions. And, as then, it will have important implications for undergraduate—and even secondary —education as well.

To cope with the complexities and ambivalences of the current period, more will be needed by the new type of generalist than professional training in the social sciences and related analytical techniques. To choose and integrate data from diverse disciplines and to orchestrate them in meaningful and operationally useful forms, the trained generalist will also require a valid standard by which to judge the consistency of policies and actions with the values and norms of behavior of the societies involved. This empathic capability is as essential for the pragmatic effectiveness of policy choices as are their technical characteristics. The requisite comprehensive conceptual framework, or way of thinking about the nature and functioning of society and culture, is derived from understanding not only of the sciences but also of the humanities. Indeed, since the essence of the humanities is the expression and the study of a society's changing sense of identity, meaning and destiny, they can play a crucial role in determining whether an integrated framework of organizing concepts is simply a set of abstract ideas or is effectively related to the on-going life of a people in all of its dimensional richness and historical continuity.[15]

This process of training the new kind of generalists and developing the new ways of perceiving and construing reality will be facilitated and in part shaped by the forthcoming generations of more sophisticated computers. For, the trained generalists will require data on so vast a scale and with such ease, rapidity and flexibility of access that only the capabilities of computers and related storage and retrieval equipment could meet their needs. Hence, it is possible that a key element in the education, as well as in the subsequent professional career, of each generalist would be a permanent association with a large versatile computer, whose programming and data-bank would be continuously developed by man-machine interactions. In this way, the information available to the policy-oriented integrator would be immeasurably increased and could be extended in accordance with his interests and requirements. More important, his continuous programming of the computer to reflect the persistent and changing interconnections, juxtapositions, perspectives and incongruities most congenial to him would enhance his creative ability to generate new and more fruitful insights and foresights.

The last sentence was worded to stress the fact that creativity is a function of the man and not of the machine, however important an aid and stimulus to his work the computer may become.[16] Some technocratic positivists, especially the more single-minded technological futurists, elevate the computer into the creative redeemer capable of remedying the deficiencies of human nature and society that men have failed to overcome. Not only can the computer store and process immensely larger amounts of data than the human mind, but also, they believe, it does so more rationally—that is, in accordance solely with the rules of logic, free of human desires and passions, beyond good and evil. Hence, they trust, it is independent of the distorting influences of both egoistic drives and sociocultural conditions and can solve problems purely by rational calculation. In this expectation, the computer becomes for the technocratic society the ultimate embodiment of the Protestant ethic, the completely impersonalized fulfillment of the worldly asceticism of the Reformation.

The Fortunes of the West

The possible 21st-century society projected in this chapter is, after all, simply an extrapolation of certain already manifest or incipient trends and of their likely interactions. Its probability is directly proportional to the extent to which these trends persist and are modified in the projected ways; or, more important, it is inversely proportional to the degree to which other trends, existing or new, become determinatively significant, and unforeseen changes occur in them as well as in the projected trends.

The latter way of stating the relationship is more useful because deliberate, conscious decisions during the intervening decades are among the major factors which could transform the course of development so substantially that the society actually existing in the next century would be quite different from all of those sketched in this chapter. And, at the risk of annoying the reader, I feel I must stress once again that the effectiveness of policy choices and program implementation are dependent on the capacity of the public and private decision makers, as well as of opinion leaders generally, to comprehend the realities with which they are trying to deal and the kinds of measures required to affect them in the desired ways. Their ability to shape the future development of their society will be largely determined by how well they understand and can orchestrate not only considerations of rational interest but also the motivational force of egoistic drives, the constraints of institutions, and the perceptual and conceptual biases of the culture.

Throughout the book, I have repeatedly pointed to the shortcomings of technocratic positivism and redemptive activism not to gratify a critical impulse but because so much depends upon them today and will even more in the years to come. Both the method of knowledge and the moral imperatives which they embody are indispensable not solely for greater social improvement and personal fulfillment in the future; they are essential for the preservation of those benefits and satisfactions that our society already provides. It is precisely because reason is the only sure cognitive instrument for controlling our destiny that I have felt it necessary to emphasize the harm that results from simplistic rationalism, technical panaceas and utopian expectations. The dreams of reason, even when shaped by systems analysis and stored in a computer's memory, are no less insubstantial than the fantasies generated without electronic assistance by primary process. And, the exaggerated self-confidence and intellectual self-righteousness to which technocratic positivists are prone are none other than modern versions of the perennial sin called "arrogance" in the Bible. The safeguard against the abuses of reason and the sense of mission is not to refrain from action because we know so little about societies and cultures and cannot avoid the dilemmas inherent in moral choice. Rather, it is to approach the inescapable task of changing institutions and behavioral norms with humility, respect and awe, with that "fear and trembling" which Kierkegaard said were the necessary prerequisites for the faithful carrying out of responsibilities.

Beyond the range over which deliberate, conscious decision making, good and bad, can influence the future course of development, the fortunes of Western society will be determined by the characteristics and trends in the sociocultural process as a whole. And, just as the creative potentialities of the human mind for increasing its knowledge of reality and for controlling its egoistic drives and selfish interests in accordance with universalistic moral standards provide grounds for a qualified optimism, at least for the longer term, so too does the nature of Western society's institutions, values and behavioral norms.

Viewed in historical perspective, Western civilization is now approaching the end of its second millennium, as it customarily dates the passage of time. Still as far as ever from reaching the Millennium for which it has yearned for the past thousand years, its accomplishments are nonetheless without parallel in human history. Although throughout the centuries it has been gravely flawed by cruelty and heartlessness, suffering incalculably from its own fanaticisms, prejudices and inordinate ambitions, and wracked by the most extreme revolutions and social convulsions, Western civilization has still been able to develop the most productive, just and self-critical society yet known on this planet. In large part, this achievement results from the original Christian fusion of the Greek passion for personal fulfillment and rational thought with the Hebrew passion for moral action. From that incongruous yet infinitely fertile union of the reasoning of Plato and Aristotle with the righteousness of Amos and the two Isaiahs have come the many and varied fruits during the millennium now ending of the ceaseless quest to understand and control the forces of nature and society, of the driving will to improve the conditions of life, and of the guilty conscience that sooner or later impels the amelioration, if not the elimination, of injustice, oppression and want.

In our own day, the institutions, values and norms of behavior that sustain and renew these motivations and capabilities are by no means weakened. On the contrary, they are stronger than ever. As the projections in this book indicate, they are likely to become even more powerful in the future. Although the potentiality for evil increases along with that for good, the record of the past and the accomplishments of the present encourage my faith that the momentum of the sociocultural process will reinforce conscious decisions and actions in enabling Western society to avoid self-destruction and even stagnation.

In other words, I cannot share the conviction of Spengler, Toynbee and other contemporary historians and philosophers, who claim that, like the late Roman Empire, we are now in the decay of our society and culture leading to the collapse of Western civilization. Nor, although the resemblances are greater, do I believe that we are in the midst of a new Hellenistic age—that incongruous period of economic expansion along with social and political turmoil, of great scientific advance with little technological application, of new redemptive religions and pessimistic philosophies, of dissatisfaction, alienation and meaninglessness —which would end in the equivalent of those two centuries of stability, order and comparative prosperity under the *Pax Romana* that lasted from Augustus to Marcus Aurelius. Rather, I find the most convincing analogy in the great transition and transformation of the 15th to the 17th centuries described in Chapter II. On the one hand, those centuries were illuminated by the onset of the age of planetary discovery; the flowering of art and literature under the inspiration of the classical revival; the first flourishing of observational and experimental science and the burgeoning of technological innovations; the welding of our existing world view; the origins of the nation-state and of rationalized elites in the governmental and economic systems; and the many other inventions and quickenings of the human spirit characteristic of the Renaissance and Reformation. On the other hand, they witnessed the disintegration of immemorial institutions and the weakening of traditional values and norms; the recurrent waves of violence and mass hysteria; the end-of-the-world pessimism, the dance of death, and the poignant sense of the sorrows of life; and the uprootedness, aimlessness and *anomie* that marked the waning of the Middle Ages.

Thus, in the last analysis, I am inclined to believe that Western society could well be on the verge of commencing a third millennium even more creative and challenging than that now ending. Certainly, as I have tried so often to emphasize, I foresee no technological utopia, no new Eden of anarchic plenty and harmless gratification of everyone's egoistic drives, no Kingdom of Heaven in the here and now. The many and grievous ills that beset us today, the unimaginable difficulties and dilemmas that lie ahead, will make the coming decades at best only somewhat less painful, frustrating and disappointing than the years now past. Yet, we will at the same time be reaping the harvest of our great heritage and realizing the unknowable potentialities of Western society's continuing dynamism and creativity.

Notes

1. By a *society* is meant a distinguishable totality of specific patterns of interrelationships among people both as individuals and as members of collectivities, that is, of institutions and organizations (for example, in modern Western societies, a family, a school, a congregation, a business firm, a government department, a trade union, a political party, a social club, etc.), which are in turn interrelated as institutional systems (such as the governmental or administrative system, the political system, the economic system, the church, the educational system, etc.). By a *culture* is meant not only the totality of distinctive material artifacts of a society and the techniques by which they are produced but also its characteristic ways of seeing, feeling, believing, aspiring and interpreting itself and the world (more technically, its values, attitudes, expectations, self-images, norms of behavior, and perceptions and conceptions of physical and social realities). These definitions are slight rephrasings of those agreed upon by the anthropologist A. L. Kroeber and the sociologist Talcott Parsons in "The Concepts of Culture and of Social System," *American Sociological Review,* Vol. 23, No. 5, October 1958, pp. 582–83. I use the term *civilization* to designate an identifiable totality of similar societies and cultures viewed in the perspective of their long historical development—for example, *Western civilization* as the common historical background and present common setting of the various contemporary forms of Western society and culture. The scope of Western civilization today embraces several groups of nations that may be regarded as constituting different sociocultural varieties of it. In addition to the North American and West European grouping, they consist of the Soviet Union and the communist states of Eastern Europe; Australia and New Zealand; the Latin American and Caribbean countries; and portions of the populations of Israel, South Africa and Rhodesia. Unfortunately, I have had to employ the adjective "sociocultural" in two senses: the first as all-inclusive in accordance with the foregoing definitions of the nouns; the second as limited to the many other institutional and cultural elements that are not included in the meaning of the terms "economic" and "political-strategic." As there seems to be no other conveniently brief word for denoting the more limited sense, I have tried to make the context indicate whether the adjective is being used in the former or the latter meaning.

2. A well-known example of the inhibiting effects of sociocultural factors on technological innovation is the absence of the wheel in the pre-Columbian New World. Although made and used in the form of calendar disks, the wheel shape was never applied for mechanical purposes even in the three most advanced New World societies (the Mayas, Incas and Aztecs). For analyses of the complex interrelationships between technological innovation and sociocultural change, see especially Lynn White, Jr., *Medieval Technology and Social Change* (London and New York: Oxford University Press, 1962), David S. Landes, *The Unbound Prometheus: Technological Change and Industrial Development in Western Europe from 1750 to the Present* (Cambridge: Cambridge University Press, 1969) and Emmanuel G. Mesthene, *Technological Change: Its Impact on Man and Society* (Cambridge: Harvard University Press, 1970).

3. Geoffrey Barraclough, *An Introduction to Contemporary History* (New York: Basic Books, Inc., 1964) Chapter I.

4. Among recent American futurist publications with a broader geographical focus which endeavor to take into account a wider variety of sociocultural factors are Herman Kahn and Anthony J. Wiener, *The Year 2000: A Framework for Speculation on the Next Thirty Years* (New York: The Macmillan Company, 1967); *Daedalus,* Vol. 96, No. 3, entitled *Toward the Year 2000: Work in Progress* (The American Academy of Arts and Sciences, Summer 1967); Daniel Bell, "Notes on the Post-Industrial Society," *The Public Interest,* Nos. 6 and 7, Winter and Spring 1967; Zbigniew Brzezinski, *Between Two Ages: America's Role in the Technetronic Era* (New York: The Viking Press, 1970). As a counterpoint to the simplistic technologism of much of the futurist literature, I recommend Lynn White, *Machina Ex Deo: Essays in the Dynamism of Western Culture* (Cambridge: The MIT Press, 1968) and Victor C. Ferkiss, *Technological Man: The Myth and the Reality* (New York: George Braziller, Inc., 1969).

5. The question of which countries to include in the Atlantic region presents some difficulties. Clearly, the larger, wealthier and more powerful nations—the United States, Germany, the United Kingdom, France and Italy—constitute its heart, and this book focuses mainly on them. However, Austria, Belgium-Luxembourg, Canada, Ireland, the Netherlands, the Scandinavian countries and Switzerland share fully in their political, economic and other sociocultural characteristics. In historical and sociocultural terms, Spain, Portugal and Greece would also have to be regarded as belonging to the Atlantic group and, in varying degree, their economies have been becoming more industrialized and integrated with the others in the past decade even though their political institutions hardly qualify as democratic. The chief difficulties relate to Japan and, to a lesser extent, Australia and New Zealand. Not only does the former lack geographical propinquity but it is also part of a different great historical tradition. Despite the effects of industrialization and the adoption of Western techniques, its society and

culture diverge significantly from those of Western Europe and North America. Nevertheless, Japan cannot be omitted from the group when economic relationships are under discussion, and it has been a major link in the network of mutual defense arrangements centered on the United States. Although also not geographically part of the Atlantic region, Australia and New Zealand are wholly Western nations in sociocultural terms. Their economies are already at high income levels and are linked closely with those of Western Europe, North America and Japan. They, too, participate in the mutual defense arrangements. Hence, despite their location, Japan, Australia and New Zealand are organically part of the Atlantic region with respect to the particular aspects noted.

6. I have found most congenial to my own way of thinking about the nature of the social process the concept of society as a system developed in Walter Buckley, *Sociology and Modern Systems Theory* (Englewood Cliffs, N.J.: Prentice-Hall, Inc., 1967). Insofar as his systems model can be applied at the level of generality in this book, I have endeavored to do so. However, I have placed greater stress on the psychocultural aspects of the social process in the forms proposed by Benjamin Nelson, "Actors, Directors, Roles, Cues, Meanings, Identities," *The Psychoanalytic Review*, Vol. 51, No. 1, Spring 1964, pp. 135–160.

7. Benjamin Nelson, cited. I prefer Nelson's term "dramatic design" to the conventional term "ideology" which, since it was coined by Destutt de Tracy more than a century and a half ago, has been used in too many different and contradictory senses.

8. Policy applications of the analysis in this book are being made in other publications, cf. "Statement of Theodore Geiger" in *A Foreign Economic Policy for the 1970s,* Hearings before the Subcommittee on Foreign Economic Policy of the Joint Economic Committee, Part 2, Ninety First Congress, Second Session, 1970; Theodore Geiger, *Transatlantic Relations in the Prospect of an Enlarged European Community* (Washington, D.C.: National Planning Association; London: British-North American Research Association; Montreal: Private Planning Association of Canada, 1970); and *U.S. Foreign Economic Policy for the 1970s: A New Approach to New Realities* (Washington, D.C.: National Planning Association, 1971).

CHAPTER II

1. Summarized in Norman Cohn, *The Pursuit of the Millennium* (London: Secker and Warburg, 1957) p. 100.

2. See especially John Leddy Phelan, *The Millennial Kingdom of the Franciscans in the New World* (Berkeley and Los Angeles: University of

California Press, 1956) *passim;* and Theodore Geiger, *The Conflicted Relationship: The West and the Transformation of Asia, Africa and Latin America* (New York: McGraw-Hill for the Council on Foreign Relations, 1967) Chapter Six.

3. The authors of secular utopias usually refuse to accept the impossibility of realizing both simultaneously and fully the incompatible absolute ideals they seek, such as personal freedom, individual equality, social justice, communal peace, and economic plenty. In effect, their utopias are constructed either by treating one or two absolute ideals as overriding and sacrificing the others to them (e.g., assuring equality and justice by suppressing personal freedom), or by assuming that, in a utopian state, people would naturally behave with complete rationality or altruistic love and, in consequence, inconsistent values would be automatically harmonized.

4. The insertion of "each" in Protagoras' sentence accords with the interpretation of his philosophy by John Burnet, *Greek Philosophy: Part I Thales to Plato* (London: Macmillan, 1928) pp. 114–115. Whether or not Protagoras intended this meaning is, however, immaterial, as the generalization it exemplifies was characteristic of Greek rationalism.

5. For an analysis of social identification in traditional societies and of the major modern example of emerging individualization in dissolving traditional societies, see Geiger, *The Conflicted Relationship,* cited, Chapters Three and Four.

6. Max Weber, *The Protestant Ethic and the Spirit of Capitalism* (New York: Charles Scribner's Sons, 1958) p. 26.

7. Many scholars have attacked Weber for presumably reducing social causation to religious changes. A careful reading of *The Protestant Ethic* should alone have dispelled this misconception—which, indeed, Weber specifically pointed out in its concluding paragraph: "But it is, of course, not my aim to substitute for a one-sided materialistic an equally one-sided spiritualistic causal interpretation of culture and of history." (p. 183). I have read many of the other criticisms of Weber's work made by political scientists and other scholars during the past decade or so. Those reflecting data unavailable in his time are often justified. However, many are differences of interpretation and hence questions of opinion, and others seem to me to be based on inadequate knowledge of the very broad range of Weber's interests and of the many different perspectives from which he approached particular subjects in the course of his analysis. Now that an English translation is available of the complete text of his monumental *Economy and Society: An Outline of Interpretive Sociology* (New York: Bedminster Press, 1968, 3 volumes), the full richness, depth and contemporary relevance of his general conceptual framework and of many of his particular insights can be more clearly seen, validating their acceptance by Talcott Parsons, Benjamin Nelson, Robert Bellah, David Little, and

others who pioneered in their application and further development in the United States.

8. Weber, *Economy and Society,* cited, Vol. 2, p. 556.

9. Weber, *Economy and Society,* cited, Vol. 2, pp. 587–8.

10. Weber, *The Protestant Ethic,* cited, p. 69.

11. See John U. Nef, *Industry and Government in France and England, 1540–1640* (Ithaca, N.Y.: Cornell University Press, 1964).

12. See Bernard Groethuysen, *The Bourgeois: Catholicism vs. Capitalism in Eighteenth-Century France* (New York: Holt, Rinehart and Winston, 1968).

13. Weber, *Economy and Society,* cited, Vol. 3, Chapters XII and XIII.

14. It is important to keep in mind, as Schumpeter pointed out, that the theorists of mercantilism differed among themselves as well as with the practitioners, see Joseph A. Schumpeter, *History of Economic Analysis* (New York: Oxford University Press, 1954) pp. 335–338. The standard account of mercantilist practice is still that of Eli F. Heckscher, *Mercantilism* (London: George Allen and Unwin, Ltd., 1934) 2 volumes.

15. The revived use of Roman law and the development of English common law were in fact more complex phenomena than the text implies. For example, the revision and reinterpretation of the common law in the 16th and 17th centuries both reflected and fostered not only the absolutism of the Tudor and Stuart dynasties but also the tendencies toward freedom of enterprise and the liberties of the subject. For an interpretation of these complex legal developments in relation to the broader sociocultural changes of the period, see David Little, *Religion, Order, and Law: A Study in Pre-Revolutionary England* (New York: Harper and Row, 1969).

16. Of course, the other side of Calvinism and its derivatives must not be overlooked—the self-righteousness of the "saints," their hardheartedness and lack of compassion, their narrow intolerance of beliefs and behavior other than those sanctioned by worldly asceticism, and their fundamentalist insistence on Biblical revelation as the source of true knowledge about nature and man.

17. See Benjamin Nelson, "The Early Modern Revolution in Science and Philosophy" in R. S. Cohen and M. Wartofsky, editors, *Boston Studies in the Philosophy of Science,* Vol. 3 (Dordrecht, Holland: D. Reidel, 1968). The passages quoted in this section are from pages 12–13.

18. As Weber explains, all bureaucratic organizations staffed by literary, legal or technical elites are inherently rationalistic regardless of the nature of the civilization in which they develop. But, only in Western society have such elites conceived of their activities as aimed at social perfection. In China, for example, the rationalistic Confucian elites were concerned with social stability and the adjustments and rectifications needed to maintain or restore it. See Weber, *Economy and Society,* cited, Vol. 3, *passim;* Fei

Hsiao-tung, *China's Gentry: Essays on Rural-Urban Relations* (Chicago: University of Chicago Press, 1953); and Max Weber, *The Religion of China* (Glencoe, Ill.: The Free Press, 1951). In present-day China, the redemptive activism of the ruling communist elites results from the impact of the West since the early 19th century. See Geiger, *The Conflicted Relationship,* cited, p. 112 including fn. 17.

19. Marx and Engels attacked the "utopian socialists" not only for their refusal to align themselves with the proletariat in revolutionary action but also for their "fantastic picture of future society"; see Karl Marx and Friedrich Engels, *The Communist Manifesto,* with an Introduction by A.J.P. Taylor (Baltimore: Penguin Books, 1967) pp. 114–118.

20. Weber, *The Protestant Ethic,* cited, p. 16.

21. In this sense, Adam Smith, Jeremy Bentham, John Stuart Mill and other utilitarians could be considered *laissez-faire* positivists in contrast to the pessimistic expectations of David Ricardo, Thomas Malthus and other classical economists responsible for the 19th-century designation of economics as "the dismal science."

22. The unparalleled productivity of the contemporary industrialized economy stems from its size, flexibility and diversification, from its intricate and highly interdependent division of labor, from its vast mechanization and growing automation, and from its more and more standardized and efficient processes and techniques. These characteristics generate the need for sophisticated knowledge and skills to carry on the increasingly complex and delicately balanced interrelationships at the *micro* level of the separate producing and consuming units (that is, organizations and individuals) comprising the economic system and at the *macro* level of the economy as a whole. Thus, the spread of the industrial mode of production originally in manufacturing and more recently in agriculture and the service sector has both fostered and been dependent on the parallel evolution of more rationalized and impersonal organizations for managing economic activities at *micro* and *macro* levels. Developing and applying the requisite technical information and procedures, and devising and administering the efficient organizational arrangements and managerial methods at both levels are the day-to-day functions of the technocrats.

23. For an analysis of the organic, affective elements in contemporary Japanese institutions, see Chie Nakane, *Japanese Society* (Berkeley: University of California Press, 1972) and John C. Pelzel, "Japanese Kinship: A Comparison" in Maurice Freedman, editor, *Family and Kinship in Chinese Society* (Stanford: Stanford University Press, 1970). Among studies of the emerging technocratic order in Japan, I have found especially helpful the volumes in the *Studies in the Modernization of Japan* prepared by general editors Marius B. Jansen and John W. Hall for the Association of Asian Studies and published by Princeton University Press; James C. Abegglen, *The Japanese Factory* (Glencoe, Ill.: The Free Press, 1958);

Robert J. Ballon, editor, *Doing Business in Japan* (Rutland, Vermont: Charles E. Tuttle Co., Inc., 1967); Marshall E. Dimock, *The Japanese Technocracy* (New York: Walker/Weatherhill, 1968); Richard Halloran, *Japan: Images and Realities* (New York: Alfred A. Knopf, 1969); Chitoshi Yanaga, *Big Business In Japanese Politics* (New Haven: Yale University Press, 1968); M. Y. Yoshino, *Japan's Managerial System: Tradition and Innovation* (Cambridge: The MIT Press, 1968). For differing forecasts of the future of Japan, see Herman Kahn, *The Emerging Japanese Super-state: Challenge and Response* (Englewood Cliffs, N.J.: Prentice-Hall, Inc., 1970) and Zbigniew Brzezinski, *The Fragile Blossom: Crisis and Change in Japan* (New York: Harper and Row, 1972). My own views are presented in Chapters IV and VI.

24. In fact, the mainstream of contemporary philosophy has continued to flow within the broad range of rationalist schools that stretches from pragmatism and naturalism to logical positivism and linguistic analysis. Important as they are as critics of rationalism and stimulators of new insights and perspectives, phenomenological and existentialist approaches do not predominate. In contrast, nonpositivistic theologies have continued to constitute the mainstream not simply in Catholicism and Protestant fundamentalism but, with greater contemporary significance, in Barth's neo-orthodoxy, Niebuhr's Christian realism, Buber's and Tillich's varieties of religious existentialism, westernized Zen Buddhism, etc. However, they have been paralleled by various positivistic reassertions of the prophetic strand in the Judaeo-Christian tradition, such as social-gospel Christianity, reform Judaism, and the "secular city" and similar movements.

CHAPTER III

1. Nicolas Berdyaev, *The Russian Idea* (Boston: Beacon Press, 1962) pp. 8–9.

2. Berdyaev characterizes it as

... one of the most poignantly painful of histories. It embraces the struggle first against the Tartar invasion and then under the Tartar yoke, the perpetual hypertrophy of the State, the totalitarian régime of the Muscovite Tsardom, the period of sedition, the Schism, the violent character of the Petrine reform, the institution of serfdom—a most terrible ulcer in Russian life—the persecution of the Intelligentsia, the execution of the Decembrists, the brutal régime of Nicholas I, the illiteracy of the masses of the people, who were kept in darkness and fear, the inevitability of revolution to resolve the conflicts of contradictions, the violent and bloody character of the revolution, and, finally, the most terrible war in the history of the world. [Nicolas Berdyaev, *The Russian Idea,* cited, p. 5.]

3. As in all agrarian societies composed of large landholders and peasants, revolts have occurred in Russian history whenever the condition of the peasantry worsened significantly. And, as elsewhere, they tended to be

directed against local grievances and authorities rather than against the distant Tzarist regime *per se*. After the consolidation of Muscovite supremacy in the 17th century, the predominant elements in political revolts against the central government were usually ethnic and religious minorities, e.g., Cossacks, Ukrainians, Schismatics, etc. The Russian people generally have been activated to revolutionary political outbreaks, as in 1905 and 1917, when the prestige of the autocratic regime was shattered by its conspicuous incompetence in time of national crisis. Otherwise, they tend to endure stoically and patiently.

4. Pluralism may be defined as a substantial degree of political and economic decision making and self-responsible activity dispersed throughout the society rather than concentrated predominantly in the central government. The disparate or conflicting interests of particular groups and organizations are not suppressed but are constrained by a sense of the general interest and oriented by universalistic values toward the achievement of widely agreed-upon national goals and priorities.

5. Analysis of centrally planned economies is outside the scope of this book, and a brief explanation must suffice. In theory, all significant economic decisions regarding production and consumption and saving and investment are supposed to be made in a fully conscious manner by the central government in accordance with a comprehensive and detailed plan for a specified time period. In real life, however, the rationality of decision making by central planners is significantly impaired by the same factors that affect governmental and private decision making in market economies. Basic cultural influences, political pressures, doctrinal prejudices, personal ambitions, and institutional contraints suppress or distort considerations of comparative costs and benefits and the choice of means effective for achieving goals. In addition, the more complex and diversified a planned economy becomes, the more difficult it is for the central planners to obtain the information required for efficient decision making within the time necessary to be effective. The growing multitude and diversity of decisions required of the planners more and more outrun their capacity to make and coordinate them with the required rapidity even with the aid of high-speed computers and the latest information storage and retrieval equipment. Owing to these intrinsic limitations of decision making in centrally planned systems, all of the East European nations, including the Soviet Union, have been trying since the early 1960s to introduce significant aspects of the market process. In varying degree, the communist countries have begun to decentralize more decision making to individual producing units in the economy and to apply more rational measurements of their performance (e.g., costs, sales and returns on capital or sales) analogous to those used in a market system. The aim of these and other reforms is to reduce the waste of resources encouraged by the physical volume criterion, which does not sufficiently constrain producing units to conserve inputs,

tailor outputs to the specific needs of consuming units, and apply adequate quality controls.

6. The successive literary contexts in which the "last best" formulation has been used aptly illustrate the shift from God to nature and from universalism to nationalism. Milton described man as the "last and best of all God's works." Robert Burns wrote "When Nature her great master-piece design'd,/And framed her last, best work, the human mind." Lincoln referred to the United States when he said "We shall nobly save or meanly lose the last, best hope of earth."

7. Adlai Stevenson, *Major Campaign Speeches* (New York: Random House, 1953) p. 262.

8. This attitude underlies the innumerable stories and jokes about the impracticality of scientists and experts compared with the quiet competence of ordinary Americans.

9. Cordell Hull, "Bases of the Foreign Policy of the United States," *Bulletin* (Washington, D.C.: Department of State, March 25, 1944) p. 276.

10. My interpretation also differs from that of the so-called "revisionists," who maintain that the cold war originated in the efforts of American leaders to impose a new imperialism on the international system and that the Soviets were only acting defensively and would have cooperated with the United States in a postwar settlement assuring their legitimate interests. However, as indicated by the text, my own observations in the Department of State during the 1940s and early 1950s and subsequent experiences and study lead me to reject their interpretations both of Soviet conceptions and behavior and of U.S. intentions and actions.

11. The term "revolutionary" is used here in the sense of a fundamental transformation of the societies and cultures of Asia, Africa and Latin America, not in the more limited meaning of a violent overthrow of their existing social relationships and institutions. Justification of this usage will be found in Geiger, *The Conflicted Relationship,* cited, Chapter Three.

12. For an analysis of U.S. misconceptions of the process of socio-cultural change in Asia, Africa and Latin America, and of consequent conflicted attitudes and unrealistic expectations regarding it, see Geiger, *The Conflicted Relationship,* cited, pp. 38–46; 271–286.

13. Indeed, the extreme undependability of peace and justice in a state of nature (i.e., in the absence of a sovereign power capable of enforcing them) was precisely the reason given by Locke for the willingness of men to

> . . . quit this condition which, however free, is full of fears and continued dangers and . . . to join in society with others who are already united, or have a mind to unite for the mutual preservation of their lives, liberties and estates, which I call by the general name—property. . . . [This is] the great and chief end, therefore, of men uniting into commonwealths, and putting themselves under governments. . . . [John Locke, *Two Treatises on Civil Government,* Book II, Chapter 9.]

14. In recent years, analysts of American foreign policy have been devoting more attention to the factors other than the substantive considerations of rational interest that play significant roles in determining U.S. objectives and activities abroad. These institutional aspects of the process of foreign policy formation include the role of Presidential leadership and power; the State Department's organization and relationships with other government agencies; the interactions between career personnel and prominent political appointees serving temporarily in policy positions; Executive Branch relationships with the Congress and the latter's own organization, procedures and powerful personalities; the influence of the "military-industrial complex"; and the more diffuse effects of partisan politics, domestic issues, special-interest groups, opinion-leader views, and the extent of popular awareness and support. Because these and other institutional factors are becoming better known, however, I have stressed the psychocultural aspects in this chapter. They are too often dismissed by theorists of international relations as mere "rhetoric." Nevertheless, the conceptions of the nature of the international system and of U.S. objectives and responsibilites explicit or implicit in official and private rationales do express self-images, values and expectations that are significant to the people generally, as well as to the elites enunciating them. Because they operate mainly through subtle psychological processes, the effects of such cultural factors cannot usually be studied by methods that seek to find a direct or immediate correlation between specific ideas and specific actions. Reflecting and in turn helping to perpetuate and reshape the normally unconscious categories of perceiving, believing and thinking, the conceptions of national identity and purpose and the related world views reinforce, color, distort, block or displace perceptions of reality and calculations of rational interest. Closely linked to subconscious processes, they are always infused with the egoistic drive needed to convert ideas into actions, sustain morale, and justify self-interested behavior. Thus, it seems to me that the psychocultural elements are as important in determining a nation's foreign policies and external activities as the conscious calculations of rational interest and the institutional pressures and restraints.

15. For example, Senator Fulbright, a leading critic of U.S. foreign policy, concluded:

It has been my purpose in this book to suggest some ways in which we might proceed with this great work [i.e., in the author's words, "to effect a fundamental change in the nature of international relations"]. All that I have proposed in these pages—that we make ourselves the friend of social revolution, that we make our society an example of human happiness, that we go beyond simple reciprocity in the effort to reconcile hostile worlds—has been based on two major premises: first, that, at this moment in history at which the human race has become capable of destroying itself, it is not merely desirable but essential that the competitive instinct of nations be brought under control; and second, that America, as the most powerful nation, is the only nation equipped to lead the world in an effort to change the nature of its politics. [J. William Fulbright, *The Arrogance of Power* (New York: Random House, 1966) p. 256.]

16. There are various theories of international relations that seek to account for the superpowers' behavior. One identifies as an intrinsic characteristic of the nation-state an urge to exercise and increase its power, which is believed to produce a mutually escalating competition in the international system. Another points to an impulse to extend domination over wider and wider areas as inherent in the condition of being a superpower. However, it seems to me that at worst these and similar ways of formulating hypotheses about superpower behavior beg the question by offering as a cause the result they are trying to explain; and that at best they are only convenient shorthand designations for complex processes composed of identifiable cultural, social and psychological factors. In the latter case, the symbolic character of such formulations becomes apparent as soon as the relevant analytical question is asked: What self-images, values and norms of behavior impel a people to increase its national power substantially, and to exercise it in ways and for objectives that go far beyond the requirements of national interest as defined in cost/benefit terms? Although answers will differ as to the specific self-conceptions and world views believed to be significant, they will of necessity involve analysis of the sociocultural characteristics of the nations concerned.

CHAPTER IV

1. Thomas Hobbes, *Leviathan,* Chapter 13.

2. John Locke, *Two Treatises on Civil Government,* Book II, Chapter 3.

3. By definition, an international system is different from a universal imperial system (*imperium mundi*) since the former always contains at least two and commonly many more independent states whereas the latter has only a single sovereign political entity. The Roman Empire and the Chinese Empire under the Han and T'ang dynasties were thought of by their inhabitants—and by subsequent generations—as universal empires because they were immense, self-contained systems embracing all known communities regarded as civilized. Nonuniversal empires, e.g., the British Empire, were, in contrast, members of international systems.

4. See Geiger, *The Conflicted Relationship,* cited, Chapter Four.

5. A general formal model for making such assessments is presented in Morton A. Kaplan, *System and Process in International Politics* (New York: John Wiley and Sons, Inc., 1957, 1964). The question raised in the text is more specifically discussed by Kenneth N. Waltz, "The Stability of a Bipolar World," *Daedalus,* Summer 1964, pp. 881–909; Karl W. Deutsch and J. David Singer, "Multipolar Systems and International Stability," *World Politics,* April 1964, pp. 390–406; and R. N. Rosecrance, "Bipolarity,

Multipolarity, and the Future," *Journal of Conflict Resolution,* September 1966, pp. 314–327. However, their discussion deals with formal analysis at the theoretical level rather than with the complexities and ambivalences of possible real-life situations and, except for the cold-war period, the historical evidence. An analysis concerned primarily with other questions but which takes some of these real-life complexities into account in considering the relationship of multipolarity and stability is Ciro Elliott Zoppo, "Nuclear Technology, Multipolarity and International Stability," *World Politics,* July 1966, pp. 579–606.

6. It is less probable that mutual deterrence will be preserved by disarmament or even by a general arms-limitation agreement than by the continued development of increasingly more advanced nuclear and conventional offensive and defensive weapons systems. Persistence of U.S.-Soviet military competition would be fostered by the logic of mutual distrust and the interest of influential institutions and groups in both countries (the "military-industrial complex") in continued research and development, with consequent need to justify the required expenditures and pressure to apply the results in military production and construction. The major offsetting considerations would be the mounting costs of deploying successive generations of competitive nuclear offensive and defensive weapons in the face of other increasingly urgent claims on resources, especially for domestic welfare; and the pressure of other nations on them to reach arms limitation agreements. On balance, then, it seems probable that the United States and the Soviet Union will continue to develop their nuclear and conventional capabilities, even though they would from time to time agree, explicitly or tacitly, not to deploy in force certain of the more costly types of nuclear weapons systems. Nor would the greater likelihood of continued growth of military capabilities preclude marginal agreements, like the nuclear testing ban, that would mitigate the harmful effects of nuclear research and development and neutralize certain peripheral areas and means of access to one another (e.g., the seabed, the moon, outer space), or "hot-line" arrangements and direct clarifications to reduce the likelihood that war between the superpowers would result from inadequate or misunderstood communications.

7. The rational calculations that enter into such decisions have been cogently analyzed in Mancur Olson, Jr., *The Logic of Collective Action: Public Goods and the Theory of Groups* (Cambridge: Harvard University Press, 1965).

8. Unlike the physical scientists and engineers who, if anything, have tended to the opposite extreme, the great majority of social scientists seem deliberately to have avoided discussion of extraterrestrial explorations. Throughout the 1950s, despite common knowledge of large Soviet and American expenditures on the development of space rockets and satellites,

most social scientists pooh-poohed—when they did not ignore—the possibility, much less the significance, of space travel. Today, they are willing to admit the military and economic importance of the space programs but efforts to explore future developments and possibilites are usually met with embarrassment. Still, there are valid *a priori* reasons for believing that other forms of intelligent life exist—although almost certainly not in our solar system—and are likely to make contact with us before we are able to master interstellar travel and reach them. Astronomers generally are convinced that there are a large number of earth-type planets in the galaxy. Probability theory would rate quite low the chance that the development of intelligent life on this planet is a unique event in the universe (*pace* the Judaeo-Christian tradition). Granted that intelligent races exist elsewhere, it is more likely that some have technologically more advanced civilizations than that ours is the most advanced. Again, normal statistical distribution would give the highest probability to a rating of average for our science and technology. Hence, if we are already capable of putting men on the moon and exploring the planets with space probes, other civilizations are likely already to be able, or might be within the next century or so, to disprove, transcend or circumvent the Einsteinian limit of the speed of light. The fact that our sun is located about halfway out from the galactic center in a region of a spiral arm rather thinly populated with stars would tend to lower the probability of early discovery by a technologically more advanced race. And, this is just as well. For, such a spacefaring race is likely to be dynamic and aggressive, if not necessarily hostile, and its impact on us could be analogous to that of the technologically more advanced and adventurous Europeans on the inhabitants of the Americas, Asia and Africa during the age of discovery. Those who yearn for the coming of a benevolent race of superior beings to save us from our follies would be likely to be sadly disappointed. But, the arrival of possibly dangerous visitors from other star systems might be the catalyst that could unite the human race.

9. The fact that the significance of the United Nations—as distinct from its functional agencies and activities—is and will continue to be in large part symbolic should not be regarded as belittling the organization. Such ecumenical symbols can have great power even long after the institutions that generated them have passed away—witness the influence of the interrelated traditions of the universal Empire and the unified community of Christendom, derived from Antiquity and the Middle Ages, in keeping alive the concept of a unified Europe throughout the five hundred years of dynastic consolidations, religious conflicts, and rise of the nation-state. As the first planetwide political institution to prefigure a world community, the United Nations exerts its symbolic power through hope rather than nostalgia.

CHAPTER V

1. The analysis in this section of the postwar movements for European union and Atlantic partnership is derived from the author's participation in or direct observation of these developments during his years of service in the Department of State and the Economic Cooperation Administration, and later as an adviser on European and Atlantic regional affairs. It seems unnecessary, therefore, to cite accounts in secondary sources.

2. This hypothesis and the supporting analysis were developed by Harold van B. Cleveland and the author in several policy papers that were instrumental in bringing about the changes in U.S. policy toward European unification outlined in the text. They were declassified only in 1970 and are available in the ECA Policy Series at the National Archives, Washington, D. C.

3. Theodore Geiger and H. van B. Cleveland, *Making Western Europe Defensible* (Washington, D. C.: National Planning Association, 1951) pp. 43–44.

4. Despite its wide currency, the term "Atlantic Community" is not used in this book. Not only the advocates of Atlantic union but also government officials and publicists in the United States, Canada and Western Europe tend to refer to the Atlantic Community as though it were analogous to the European Community. In consequence, the term implies a greater range of common interests and a more institutionalized structure for expressing them than now exist in the Atlantic region or are likely to evolve.

5. The theory of functional integration was first propounded by the political scientist David Mitrany during the 1930s as a means of bringing about a new peaceful system of world order. David Mitrany, *A Working Peace System* (London: Royal Institute of International Affairs, 1943). However, it was first applied to the problems of European unification in the postwar period by Jean Monnet and his followers. In the form of neo-functionalism, the theory has been further developed by Ernst B. Haas, *Beyond the Nation-State: Functionalism and International Organization* (Stanford: Stanford University Press, 1964) Part I, and in the new preface to the 1968 edition of his *The Uniting of Europe* (Stanford: Stanford University Press).

6. For an analysis of the tensions between these two conceptions and a critique of official and other efforts to reconcile them, see the cogent analysis in Harold van B. Cleveland, *The Atlantic Idea and Its European Rivals* (New York: McGraw-Hill for the Council on Foreign Relations, 1966).

7. Under Article 24 of the General Agreement on Tariffs and Trade, member countries are permitted to form two types of free-trade arrangements: a customs union and a free-trade area. In both, substantially all

trade among the participants must be free of tariffs and quantitative restrictions, and those against nonparticipants must not be greater under the arrangement than they were before. The difference between a customs union and a free-trade area is in the degree of economic integration involved. In a customs union, the members adopt a common external tariff against nonmembers which, in turn, necessitates a common foreign-trade policy and the coordination of other national economic policies significantly affecting the maintenance of free trade. In a free-trade area, members retain their own systems of trade restrictions against nonmembers and do not, therefore, have to coordinate their national economic policies as closely.

8. Europeanists maintain that had de Gaulle not come to power, or had he done so five years later, the unification movement would have advanced far enough to be unstoppable. As the analysis in this chapter indicates, I believe that the present extent and future prospects of European unification would not be fundamentally different from what they are had either of these two contingencies happened. Institutional and attitudinal changes as profound as those involved in politicoeconomic unification are likely to be influenced in detail and in timing by the actions of a prepotent leader, but his decisions cannot make the decisive difference as to whether or not the process occurs.

9. My data on and interpretation of European attitudes are in accord with those of Karl W. Deutsch, Lewis J. Edinger, Ray C. Macridis and Richard L. Merritt, *France, Germany and the Western Alliance: A Study of Elite Attitudes on European Integration and World Politics* (New York: Charles Scribner's Sons, 1967). They also agree generally with those of Daniel Lerner and Morton Gorden, *Euratlantica: Changing Perspectives of the European Elites* (Cambridge, Mass.: The M.I.T. Press, 1969), although these scholars regard the prospects for European union as more favorable. A difference in methodology may account for the different assessment. My interviews were not conducted with standard questionnaires but were adapted to the interests, thought processes, and willingness to communicate of the persons involved. While this approach does not yield data that can be analyzed statistically, it also does not abstract from the complexities of real-life situations. I find it preferable because it reveals much better the ambivalences and ambiguities of actual attitudes and opinions and the relative intensities with which they are held.

10. Most opinion polls testing popular attitudes on European union have been simple "for or against" questionnaires. Since the late 1940s, they have invariably yielded a substantially higher percentage of positive than of negative responses. Only a few efforts have been made to ascertain the relative intensity of popular attitudes by asking whether the respondents would be willing, for example, to pay slightly higher taxes or to accept small declines or forgo some future increases in living standards for the

sake of European union. In polls of the latter type, the favorable responses fall to well under 50 percent.

11. In the early 1950s, the young technocrats were eager to work under Monnet in the newly established European Coal and Steel Community. Again, during the EC's exciting formative years in the late 1950s and early 1960s, they sought and obtained positions in the secretariat in Brussels. However, as the EC's momentum slowed in the mid-1960s, many— though by no means all—drifted back to their countries to work in the expanding national bureaucracies and private sectors. Whether they will once again prefer Brussels depends upon whether that is "where the action is."

12. Also, the technocrats give higher priorities than the people generally to accelerating the development of the electronic, nuclear, aerospace and other science-based industries. But, unlike the "gap" publicists, they do so primarily because these industries are important for modernizing their societies rather than because they are also essential for achieving super-power status. Because of their own better understanding of the nature of technological disparities and the means for narrowing them, few technocrats were to be found in the ranks of the Europeans most alarmed about and most vociferous in publicizing the technological gap and its presumed consequences for European independence and influence. A representative technocratic view is that of one of Europe's leading scientist-industrialists, Professor Dr. H. B. Casimir, then Managing Director of N. V. Philips Gloeilampenfabrieken, "Science and Technology in the U.S. and in Europe," *Economic Quarterly Review* (Amsterdam-Rotterdam Bank N. V.) No. 14, September 1968, pp. 13–17. In contrast, for the opinions of a prominent European businessman deeply alarmed by the technological gap, see Aurelio Peccei, *The Chasm Ahead* (New York: The Macmillan Company, 1969).

13. It is important to distinguish between the prewar and postwar types of strongly nationalistic Europeans. The former were inclined to be aggressive, imperialist, chauvinist, xenophobic and racist; these objectionable traits tend to be either absent or largely latent in the latter. Remnants of the prewar type can be found in the ultranationalist and extreme rightwing parties on the continent.

14. The fact that the EFTA did not have free trade in agricultural products is not a valid objection to this assessment of its accomplishments. True, the problems of negotiating the abolition of agricultural trade barriers would have been exceedingly difficult for the EFTA's members. But, once accomplished, such a development would not have necessitated substantial supranational authority in the EFTA—no more so than it has in the Community. Under the EC's common agricultural policy, major policy changes must be approved by the Council of Ministers and the funds administered by the Commission are allocated to it by national governments.

The amount of supranational authority exercised by the Commission in connection with the common agricultural policy is in fact quite small.

15. The risk of withdrawing the American guarantee is that, depending on circumstances, the Europeans would react not by developing a credible deterrent of their own but by concluding either that the Soviet danger was past, in which case they needed none, or that their only hope of preserving some measure of independence was by appeasing the Russians, thus becoming Soviet client states. The latter possibility is called "Finlandization" in Western Europe.

16. In his press conference of January 21, 1971, President Georges Pompidou sketched a course of development for the Community leading to a "European confederation." He envisaged member governments appointing "European ministers," who would constitute a collective executive to discharge by unanimous agreement "European responsibilities" with the assistance of "specialized bodies" for planning and implementation and under the supervision of a "real European parliament." The meanings of none of these terms were defined.

Chapter VI

1. Hostile attitudes also enter into this European tendency to judge the United States more severely than the Soviet Union, as exemplified by the jealousy of the older type of European nationalist at his country's loss of great power status, or the propensity of many European liberals and socialists to believe that "Wall Street imperialists" make U.S. foreign policy.

2. Overt hostility to the United States does characterize a minority of European elites whose feelings are most nationalistic. They are anti-American because they regard the wealth and power, the NATO "hegemony"— loose as it has become—and the sense of mission of the United States as the most immediate and serious external impediments to the achievement of superpower status by their own nations individually or, as the case may be, as the dominating member of a united Europe. Such views do not make them pro-Soviet, although they do provide the basis for the kind of implicit informal cooperation that existed between General de Gaulle and the French Communist Party.

3. Whether exports have stimulated growth, or growth stimulated exports, or both, or neither, and in what circumstances and degree such interrelationships have existed in the Atlantic region since World War II are still controversial questions among economists, and some evidence can be found to support each contention. For an excellent brief critique of various theories of economic growth in the Atlantic countries, see Charles P. Kindleberger, *Europe's Postwar Growth, The Role of Labor Supply* (Cambridge: Harvard University Press, 1967).

4. The magnitude of the output of U.S. subsidiaries abroad has been estimated by Judd Polk at approximately $140 billion a year, of which ⅔ is in Canada and Western Europe. See the unpublished paper by Judd Polk, "World Companies and the New World Economy," U.S. Council of the International Chamber of Commerce, March 1971.

5. Impressive evidence of these developments can be found in Dan Smith, *Management: Europe Wakes Up* (London: *The Economist,* Brief 21, 1970).

6. Considering the economic and political importance of the balances of payments of Atlantic countries and of the national policies designed to affect them, the deficiencies of this form of national economic accounting are regrettable, to say the least. Neither the identification and definition of the component parts of the balance of payments, nor the concepts of surplus and deficit, nor the interpretation of their significance are as clear, objectively determined, and free of unrecognized sociocultural influences as the uses made of them would warrant. Their clarity and reference to objective realities are impaired not only by the arbitrariness of some of the definitions but also by the fact that several important components are estimated indirectly rather than measured directly. Attitudes toward borrowing and lending, thrift and saving, abstinence and consumption, self-punishment and self-indulgence derived from the cultural traditions surveyed in Chapter II play significant, though largely unconscious, roles in both American and European interpretations of the balance of payments and its implications for national policies. The technical perplexities of definition and interpretation are analyzed in C.P. Kindleberger, "Measuring Equilibrium in the Balance of Payments," *Journal of Political Economy,* Vol. 77, No. 6, Nov./Dec. 1969, pp. 873–891.

7. The international monetary system is worldwide in scope, embracing more than a hundred nations. Yet, its nature, direction of evolution, and condition at any given time are almost completely determined by the relationships among the leading countries of the Atlantic region. Their power is exercised through a caucus known as the Group of Ten, consisting of Belgium, Canada, France, Germany, Italy, Japan, the Netherlands, Sweden, the United Kingdom and the United States, with Switzerland cooperating informally. The Group of Ten is in virtually continuous consultation either under its own auspices or through the intergovernmental monetary and economic institutions of the Atlantic region, particularly the IMF itself, the OECD, and the monthly meetings of central bankers at the Bank for International Settlements in Basle. In consequence of its domination by the Atlantic nations, the international monetary system has not been, and is not likely soon to be, fully responsive to the needs and limitations of most Asian, African and Latin American countries. The latter's interests were recognized for the first time in 1972, when the Group of Ten reluctantly agreed to allow 10 representatives of these nations to participate in the

discussions of reforms in the international monetary system. The fact that the Group of Ten will nevertheless continue to dominate decision making on this subject validates treating the international monetary system as though it were equivalent to the regional monetary system of the Atlantic nations.

8. *International Financial Statistics* (International Monetary Fund, June 1972) Vol. XXV, No. 6, pages 19 and 23. The estimate of the percentage of dollars in international monetary reserves for each of the years given is based on (1) the amount of U.S. liquid liabilities to official foreign creditors reported by the U.S. government, plus (2) a portion of the difference between, on the one hand, the sum of these U.S. liabilities and of the U.K. sterling liabilities to foreign central monetary institutions reported by the British government and, on the other, the sum of foreign-exchange reserves reported by the monetary authorities of all the holding countries. The IMF explains (p. 45) that the difference consists of inconsistencies in statistical reporting and of foreign-exchange reserves, including Eurocurrency deposits, held in forms other than the liabilities reported by the United States and the United Kingdom. Although the IMF makes no estimate of the dollar portion of the difference, an indication of its magnitude can be inferred from the published statistics, which gave the total difference as $14 billion at the end of 1970 (*International Financial Statistics,* cited, p. 23). In its 1971 *Annual Report* (p. 23), the IMF stated that a survey of 51 central banks revealed that, at the end of 1970, they held $9 billion of reserve dollars outside the United States, i.e., in forms other than those reported by the U.S. monetary authorities. Accordingly, I have used the percentage of ⅔ in estimating the dollar portion of the total difference for 1970, as well as for 1965 when the figure was so small as to be negligible. However, it is more difficult to estimate the dollar portion of the difference at the end of 1971 for two reasons. First, in June 1971, the leading European central banks agreed temporarily not to invest additional amounts of reserve dollars in the Eurocurrency market and, second, they and other central banks sought to limit their mounting accumulations of dollars during the prolonged crisis of that year by converting some of them into other reserve assets. My interviews with national and international monetary authorities lead me to conclude that the share of dollars in the total difference of $19 billion at the end of 1971 was probably around a half.

9. Long-range balance-of-payments forecasts are notoriously unreliable. One reason is the definitional and conceptual weaknesses noted in footnote 6. Another is the fact that forecasts based on economic trends alone can be seriously wrong in consequence of unanticipated political developments, as were the predictions made in the early 1960s, prior to escalation of American involvement in Indochina, regarding the prospective disappearance of the U.S. deficit. And, even if noneconomic factors are not determinative,

long-range forecasts can be mistaken, as was the widely accepted prediction during the postwar period that a U.S. surplus was inherent in the structural relationships between the American economy and the rest of the world and would persist indefinitely. For these reasons, a definitive judgment cannot be made on the basis of current long-term forecasts of the U.S. balance of payments.

10. The doctrine of graduated response envisages that fighting could begin between the Soviet Union and the United States involving the use only of nonnuclear weapons, and that its scale might be increased by steps into the tactical and strategic nuclear ranges, thereby allowing time for negotiations to settle the dispute before the two nations were irreversibly committed to full-scale nuclear war. As explained in Chapter IV, this possibility is certainly relevant to the kinds of situations that might arise in Asia, Africa and Latin America, where Soviet and U.S. conventional forces might intervene in local conflicts and become engaged with one another. But, it is hard to imagine circumstances developing in Western Europe in which the doctrine would be applicable unless the U.S. nuclear guarantee ceased to be credible either to the Russians or the West Europeans. In the absence of such a change, the Soviets would know that, if they moved their troops westward to support a communist regime, as discussed in Chapter IV, or to seize Berlin, the United States would be prepared to respond within a few days by nuclear means. Hence, if the Soviets had decided upon such a move, they would be most likely to strike first at the United States with nuclear weapons. In other words, the Russians would not make the decision to move their troops unless they were resolved on a world nuclear war, in which case they would launch it to begin with. In part, the graduated-response doctrine was adopted in the early 1960s to counter the West Europeans' fear that their territories could be defended only by nuclear means. However, in recent years, I have found fewer and fewer Europeans who believe that the doctrine is relevant to the kinds of conflict situations likely to arise in Europe.

CHAPTER VII

1. In the long history of science, the method of analogy has certainly been applied on many occasions to yield absurd results. Nevertheless, it would be equally absurd to throw out the baby with the bath water. Properly used, it is an essential technique of scientific analysis, and is especially valuable when direct methods of observation and verification are unavailable.

2. Societies are particularistic when their constituent social units, institutions and groups struggle among themselves to advance or protect their

own interests with little concern for the general interest of the society as a whole. Particularism is most pronounced during the transformation of organic traditional societies, as in Europe during the late medieval and early modern periods and in Asia and Africa during the 20th century. For an analysis of the causes and consequences of particularism in contemporary Asian, African and Latin American societies, see Geiger, *The Conflicted Relationship*, cited, *passim*.

3. This process would be analogous to the emergence of the *patronate* under similar conditions during the last century of the Roman Empire's decline. In this relationship, locally powerful landowners took under their protection and control smaller proprietors and others threatened by the disorders of the times in return for their personal loyalty and economic support.

4. Sigmund Freud, *Civilization and Its Discontents* (Garden City, N.Y.: Doubleday Anchor Books, 1958) p. 47.

5. Freud, cited, pp. 21–22 fn.

6. On the role of play, see J. Huizinga, *Homo Ludens: A Study of the Play-Element in Culture* (London: Routledge and Kegan Paul, Ltd., 1949).

7. Douglas H. Heath, *Growing Up in College: Liberal Education and Maturity* (San Francisco: Jossey-Bass, Inc., 1968) p. 188.

8. See the comparison between institution-reared and family-reared children in Bruno Bettelheim, *The Children of the Dream: Communal Child-Rearing and American Education* (New York: The Macmillan Company, 1969) *passim*.

9. Bettelheim, cited, p. 320.

10. Some futurists envisage that technological advances will decisively reinforce the effects of economic interdependence and the common threats of nuclear destruction or ecological disaster in leading to world unity. This is another example of the over-valuation of technology's determinative power. It would certainly facilitate world unification by providing the technical means for greater integration. But, whether technological advances would be used for this purpose or not depends on whether nations and peoples want them to be. The likelihood that they will in the foreseeable future was assessed in the final section of Chapter IV.

11. Max Weber, *The Protestant Ethic and the Spirit of Capitalism,* cited, p. 182.

12. Freud, cited, p. 103.

13. See, for example, Theodore Roszak, *The Making of a Counter Culture: Reflections on the Technocratic Society and Its Youthful Opposition* (Garden City, N.Y.: Doubleday Anchor Books, 1969), which sympathetically but critically analyzes the efforts of young people to create a viable alternative to the technocratic society.

14. Their conceptual biases are in part responsible for the fact that economics is the most influential of the social sciences in policy making

and program implementation. The preponderant role of economics is, of course, owed in part to the fact that its proper subject is the production and distribution of goods and services, that is, of the resources essential for sustaining human life and meeting many of the most important individual and social goals. But, it is also the most rationalistic of the social sciences and, hence, its principles and methods tend to be congenial to the professionalized elites of an increasingly technocratic society. The analytical process of abstracting from the complexities and contradictions of reality and constructing logically consistent sets of generalizations is more satisfying in economics than in other social sciences because, by the nature of the phenomena with which it is concerned, it has much greater scope for significant quantification and the expression of relationships in mathematical forms. Sociology and political science, for example, deal primarily with interactions among the persons comprising institutions and groups. These relationships are difficult to measure and express mathematically, and those aspects of sociological and political phenomena that can be treated quantitatively—e.g., demographic data, voting behavior, attitude tests, opinion polls, etc.—are in some cases too gross to reveal the most significant factors subsumed in them or they illuminate those of only peripheral importance. True, as explained in Chapter III, the other social sciences have also been flourishing in the United States in the mid-century decades. Nonetheless, their role in national policy making has been much smaller than that of economics not only for the obvious first reason noted above but also because they are believed to be less "scientific" since they deal with aspects of the social process that are neither quantifiable nor susceptible to interpretation solely in terms of rational calculation, of the conscious weighing of costs, benefits and risks.

15. This added empathic quality that knowledge of the humanities brings can help properly trained generalists to counteract the tendency toward psychological (not philosophical) Platonism on the part of many social scientists—that is, to regard their abstract models as "more real" than the phenomena they are supposed to represent. Although they are usually careful to specify in advance the limitations of their models, the social pressures on specialists to propose solutions to real-life problems and the emotional investment in their own work often tempt them to deduce policy prescriptions without regard to their empirical validity, as though the course of events would naturally conform to the idealized relationships embodied in the models. Assuming that he has the requisite personality, the more a properly trained generalist understands and can utilize the insights and experiences subsumed in the humanities, the better will he be able to offset this unconscious Platonism in selecting, integrating and translating into practicable prescriptions the knowledge derived from the various sciences and technologies.

16. Some positivists assert that the creativity of the computer is demon-

strated by the fact that it can design progressively more sophisticated generations of computers. But, all that this capability shows is that it can handle much more complex design problems than can human engineers. It is still the latter who specify the characteristics of the new computers to be designed, as well as supplying the will to initiate the design process and to use its results.

Index

currency, 201, 204, 207, 209, 215
customs union, 288–89

decline, possible future, 226–31
defense treaty systems, 90–91
definition of terms used in this book, 275
De Gasperi, Alcide, 138
de Gaulle, Charles; aided by French technocracy, 47; and European unification, 143, 146, 152, 153, 159, 164; mentioned, 198, 220; and nuclear force, 100; provocation of U.S. by, 101, 112
Descartes, René, 31
deterrence: graduated, 221, 294; mutual, 286
dissent: in a future society, 253–55, 260–62; youthful radical, 266–67
"dramatic design," 10, 58, 277
Dulles, John Foster, 138
Durkheim, Emile, 78

East Europe, Soviet alliances in, 97–98
EC. See European Community
ECA (Economic Cooperation Administration), 137
ecological threat, as leading to a world community, 127
economic aspects of Atlantic system: balance problems, 197–202; bloc formation, 110, 213–18; ECA, 137; EEC, 129, 140–41; EFTA, 141–47, 156–57, 182, 212; and future of EC, 165–69; future prospects for, 218–24; growth's importance, 50–55; integration, 130, 139–43, 155–65, 179–90; integration vs. unification, 155–65; interdependencies among nations, 7–8; Marshall Plan, 131, 137, 139, 171–72, 181; monetary systems, 202–12; multinational companies, 184–87; OECD, 180, 208, 215; OEEC, 139–40, 180; postwar problems, 131–32; production integration, 183–84, 186–187; tripolarization of, 210–18; see also European Community
Economic Cooperation Administration, 137
economics: influence on policy making, 295–96; macro vs. micro systems, 280; from patrimonial to liberal to technocratic management of, 41–45; positivism's effect on, 77–79; and rationalism, 23–25; and

resource allocation, 51–55, 108, 197–98; Soviet policy on, 67–68; in a technocratic society, 36–39
ECSC (European Coal and Steel Community), 140–41
EDC (European Defense Community), 140
EEC (European Economic Community), 129, 140–41
EFTA (European Free Trade Association), 141–47, 156–57, 182, 212
Eisenhower, Dwight D., 138
elites, technocratic, 247–50
empires, universal, 285
Engels, Friedrich, 35, 40
enterprises: internationalization of, 216–17; multinational, 182–87, 257; rise of modern, 36–39
EPC (European Political Community), 140
ERP. See Marshall Plan
Euratom, 140–41
Eurocurrency, 182, 186
European Atomic Energy Community, 140–41
European Coal and Steel Community, 140–41
European Community, 141–69: achievements and failures, 145, 146; ambivalence on economic integration, 189; economic power and influence of, 173–74; and integration vs. unification, 218–19; Mansholt Plan, 195, 197; and multinational enterprise, 184; monetary union, 147–48; to become a proto-superpower? 109–12; see also European unification
European Defense Community, 140
European Economic Community, 129, 140–41
European Free Trade Association, 141–47, 156–57, 182, 212
European Movement, 134
European Parliament, 158–60
European Political Community, 140
European Recovery Program. See Marshall Plan
European unification, 129–69: attitudes toward, 148–55, 289–90; de Gaulle's influence on, 289; functional vs. constitutional approach to, 134–35; future of, 165–69; compared with integration of Europe, 155–56; in new nationalism period, 143–48; opinion polls on, 289–90; political perplexities of, 158–61; in postwar period, 130–43; and U.S.